S0-BLA-252

TRACE METALS IN THE ENVIRONMENT

VOLUME 4—Palladium & Osmium

by

Ivan C. Smith
Senior Advisor for Environmental Science

Bonnie L. Carson
Associate Chemist

Thomas L. Ferguson
Principal Chemical Engineer

Midwest Research Institute
Kansas City, Missouri

Published 1978 by Ann Arbor Science Publishers, Inc.
P.O. Box 1425, Ann Arbor, Michigan

Library of Congress Catalog Card Number 77-088486
ISBN 0-250-40217-3
Manufactured in the United States of America

The work upon which this publication is based was performed pursuant to Contract No. N01-ES-2-2090 with the National Institute of Environmental Health Sciences, Department of Health, Education, and Welfare.

PREFACE

In 1972, the National Institute of Environmental Health Sciences initiated a program with Midwest Research Institute to assemble information on production, usage, natural environmental levels, anthropogenic sources, human and animal health effects, and environmental impacts of selected trace elements. This book on palladium and osmium is one of a series of comprehensive documents which has resulted from that program. Scientists in many disciplines should find this work to be of value.

The authors wish to express their appreciation to Dr. Ernest Angino of the University of Kansas; Dr. Thomas O'Keefe and Dr. Arthur Morriss of the University of Missouri, Rolla; Dr. Roscoe Ellis of Kansas State University; and to our colleague at Midwest Research Institute, Mrs. Connie Weis, for assistance with this book.

The authors wish to express their thanks to Dr. Harold Orel for editorial comments and to Miss Audene Cook and her staff for typing the manuscript. We also wish to acknowledge the information provided us by Mr. Howard Martin of Martin Metals, Inc.; Mr. David Luning of Matthey Bishop, Inc.; Mr. Larry Greenspan and Mr. N. F. Carsillo of Engelhard Industries Division; Mr. A. Dipiazzo, Goldsmith Brothers, Division of National Lead; Dr. William D. Balgord of the New York State Environmental Conservation Department; Dr. Bruce Gonser, retired, of Columbus, Ohio; Dr. Willard Hoehn of G. D. Searle Company; and numerous other spokesmen for industry and government regulatory agencies.

Finally, we especially acknowledge the support, encouragement, and patience of Dr. Warren T. Piver (project officer) and Dr. Hans Falk of the National Institute of Environmental Health Sciences.

PREFACE

In 1974, the National Institute of Environmental Health Sciences initiated a program with the Midwest Research Institute [illegible] information [illegible] human and animal health [illegible]. This book [illegible] Scientists [illegible] should find this work to be of value.

The authors wish to express their appreciation to [illegible] Kansas State University [illegible] University of Missouri [illegible] Kansas State University [illegible] for assistance with [illegible].

The authors also wish to express their thanks to [illegible] Environmental Protection Agency [illegible] National Institute [illegible] Dr. William [illegible] Department [illegible] Dr. Milford [illegible] industry and government regulatory agencies [illegible].

Finally, we especially acknowledge the support, encouragement, and patience of Dr. [illegible] Project Officer, and Dr. [illegible] of the National Institute of Environmental Health Sciences.

PALLADIUM

TABLE OF CONTENTS

		Page
Summary		3
I.	Introduction	6
II.	Sources of Palladium	6
	A. Natural Sources	7
	B. Metal Scrap	8
III.	Economics	9
	A. U.S. Supply and Demand	9
	B. Worldwide Sources for U.S. Consumption	10
IV.	Processing of Palladium Ore, Metal, and Scrap	12
	A. Mining	12
	B. Refining	13
	C. Secondary Recovery	20
	D. Metal Processing	27
	1. Palladium metal and alloys	28
	2. Palladium coatings	29
V.	Uses of Palladium	32
	A. Noncatalytic Uses	33
	B. Catalytic Uses	34
	1. Hydrogenation in the chemical industry and research	34
	2. Hydrogenation in the pharmaceutical industry	35
	3. Petroleum refining	36
	4. Catalytic mufflers	36
	5. Catalytic air pollution control other than catalytic mufflers	42
	6. Miscellaneous catalyst uses	42
VI.	Losses to the Environment	43
	A. Production and Fabrication Losses	43
	B. Losses from Use and Disposal	47

TABLE OF CONTENTS (Continued)

	Page
1. Catalyst losses by petroleum refining and chemical industry	47
2. Losses to the environment from catalytic mufflers	49
3. Attrition of palladium during contact use	50
4. Dental	51
5. Losses from other uses	51
C. Summary of Losses	51
VII. Physiological Effects of Palladium and Its Compounds	52
A. Human Effects	52
B. Animal Effects	53
C. Plant Effects	55
D. Molecular Biology	55
VIII. Human Health Hazards	56
IX. Conclusions and Recommendations	57
A. Conclusions	57
B. Recommendations	58
Appendix A - Tables Cited in the Text	59
Appendix B - Physical Properties of Palladium	94
Appendix C - Chemistry	105
A. Corrosion	106
B. Oxidation and Volatilization	109
C. Organic Chemistry	110
D. Palladium Carbonyls	112
E. Alloys	112
Appendix D - Analysis	115
A. Gravimetric and Volumetric Analyses	116
B. Fire Assay	116
C. Spectrochemical and Spectrographic Methods	117
D. Polarography	118
E. Neutron Activation Analysis	118
Bibliography	119

OSMIUM

Page

Summary. 143

I. Introduction 146

II. Background 146

III. Mining and Processing. 150

A. Mining and Refining of Osmium. 150
B. Recovery for Reprocessing. 156
C. Osmium Losses During Refining. 157

IV. Consumers and Usage. 158

A. Research Uses. 160
B. Medical Uses of Osmium 161
C. Chemical Industry. 162
D. Electrical Industry. 163
E. Other Uses 163

V. Environment Effects. 164

VI. Health Hazards 167

Appendix A - Chemistry of Osmium and Its Compounds 169

Appendix B - Physical Properties of Osmium Compounds 175

References 179

Extended Osmium Bibliography 181

Subject Index 191

TABLE OF CONTENTS (Continued)

LIST OF FIGURES FOR PALLADIUM

Figure	Title	Page
1	Recovery of Platinum Metals from the Sudbury Ores	14
2	Palladium Recovery Process.	15
3	Typical Copper Smelting and Electrolytic Refining Practice.	17
4	Typical Lead Smelting and Refining Practice	19
5	Smelting and Refining of Secondary Precious-Metal Scrap . .	21
6	Secondary Recovery by a Copper Refiner.	22
7	Silver Refinery Flowsheet	23

LIST OF TABLES FOR PALLADIUM

Table	Title	Page
A-I	Concentration of Palladium in Various Sources	61
A-II	Natural Alloys and Intermetallic Compounds of Palladium . .	62
A-III	Natural Palladium Compounds with Oxygen, Sulfur, Tellurium, Arsenic, Antimony, Bismuth and Tin	63
A-IV	Imports of Refined Palladium.	64
A-V	Imports for Consumption of Palladium, Unwrought: Partly Worked; or as Bars, Plates, and Sheets ≥ 0.125 in. Thick	65
A-VI	U.S. Refinery Production of Palladium, Troy Ounce	66
A-VII	General Services Administration Stock of Palladium.	67
A-VIII	U.S. Palladium Stocks at Year End, Troy Ounce	68

TABLE OF CONTENTS (Concluded)

LIST OF TABLES FOR PALLADIUM (Concluded)

Table	Title	Page
A-IX	U.S. Exports of all Platinum-Group Metals Except Platinum	69
A-X	Palladium Sales by Industry, Troy Ounce	70
A-XI	Palladium Supply and Demand, 1968, Troy Ounce	71
A-XII	Palladium Metal Works, USA	72
A-XIII	Noncatalytic Uses of Palladium	74
A-XIV	Catalytic Uses of Palladium	85
A-XV	1975 Catalytic Converter Systems	92
A-XVI	INCO Smelter Stack Emissions, 1971	93
B-I	Physical Properties of Palladium and Some of its Alloys	96
B-II	Physical Properties of Palladium Compounds	98
C-I	Corrosion Resistance of Palladium and Platinum	107
C-II	Corrosion in Fused Salts	108

PALLADIUM

Ivan C. Smith
Bonnie L. Carson
Thomas L. Ferguson

SUMMARY

This report summarizes available information on the natural occurrence, processing, uses, and disposal of palladium and its compounds. We have estimated losses of palladium to the environment during these activities, attempted to identify the forms that the lost palladium assumes, and assessed human-health hazards associated with environmental contamination by palladium.

The principal findings of the report may be summarized as follows:

A large fraction of the world's primary production of platinum-group metals--ruthenium, rhodium, palladium, osmium, iridium, and platinum--is recovered as a by-product of copper and nickel sulfide-ore refining. The rest is recovered from placer deposits of platinum-group metals and gold and from lead and gold ores.

The principal sources of platinum-group metals in the world are the Bushveld Complex of South Africa; the Sudbury District of Ontario, Canada; and the regions of Norilsk and the Kola Peninsula of the USSR. Minor sources are placer deposits in Alaska, Columbia, Ethiopia, Japan, Australia, and Sierra Leone.

The major domestic source of platinum-group metals is secondary recovery from metal scrap, chiefly electronics scrap. The amount of palladium refined from used sources was 161,099 troy ounces in 1971. The amount refined without transfer of ownership (toll refined) in 1971 was 593,842 troy ounces.

Most of the <u>new</u> platinum-group metals recovered in the U.S. are by-products of copper sulfide-, lead-, and gold-ore refining. Industrial contacts estimated that 75-95% of the new palladium recovered each year in the U.S. is from copper refining. Possibly as much as 15% is recovered as a by-product from lead refining. The amount of new domestic palladium refined in 1971 was 32,102 troy ounces.

Chief free-world producers of palladium include International Nickel Company of Canada, Ltd. (INCO); Falconbridge Nickel Mines, Ltd., Sudbury, Ontario; and Rustenburg Platinum Mines, Ltd., of South Africa. Most of the Canadian and South African production is refined in England. Soviet platinum metals are refined in the USSR and distributed to the U.S. directly or via European countries. Chief refiners and distributors of new palladium produced in the U.S. are the major copper refiners; Engelhard Industries, Inc.,; and Matthey Bishop, Inc.

Reported sales of palladium to industry have been assumed to closely approximate domestic consumption. The average amount of palladium sold annually to industry from 1962 to 1971 was 677,608 troy ounces. The average annual fraction used by the various industries in that period was electrical, 57.38%: chemical, 25.73%; dental and medical, 7.55%; jewelry and decorative, 2.83%; petroleum, 2.52%; glass, 0.42%; and miscellaneous and undistributed, 3.42%.

Noncatalytic uses of palladium comprise about 70% of all palladium consumed by industry. About 50% is used as electrical contacts in telephone equipment. Other applications include hydrogen-diffusion apparatus and cladding of process equipment.

Palladium is used in the chemical industry primarily for catalysts in the hydrogenation of intermediates in the production of pharmaceuticals, fragrances, pesticides, dyes, etc., and in the anthraquinone process for hydrogen peroxide production. Palladium(II) compounds are used widely for catalyzing oxidation and carbonylation reactions, especially in commercial processes for producing vinyl acetate, acrylic acid, and acetaldehyde. Supported palladium or palladium zeolite catalysts are used for hydrocracking in petroleum refining.

A major new use for palladium ($>$ 100,000 troy ounces per year) is in the catalytic muffler for oxidation of hydrocarbons and carbon monoxide and reduction of nitrogen oxides in automobile-exhaust emissions.

We estimate that the average annual loss of palladium to the domestic environment in recent years has been about 244,000 troy ounces, distributed among losses to the atmosphere, during copper smelting 1,200, losses during primary and secondary refining 44,000, petroleum refining 50-340, electrical uses 39,000, and chemical-industry catalysts 160,000 troy ounces.

Annual losses of palladium to the environment will increase with the introduction of catalytic automobile-emission-control devices. It is presently estimated that 10 million vehicles will be equipped annually with catalytic mufflers and that a muffler's average life will be approximately 5 years (50,000 miles). Using estimates of 0.01 troy ounce per converter per automobile, 40% catalyst attrition during the useful life of the catalyst, and up to half of this attrition occurring during the first year of operation, one calculates that 8,000-20,000 troy ounces of palladium would be lost from this source in 1975 and that the annual loss will approach 40,000 troy ounces by 1979. This loss is expected to occur as fine particulate palladium metal or as some unknown palladium compound. No information was found on the hazards of breathing fine palladium particulates.

The remainder of the palladium in spent mufflers will likely be discarded to the environment until such time as platinum-palladium prices or the number of spent mufflers justify recovery. Estimated value of residual platinum and palladium in spent mufflers will approach $33 million annually by 1979, based on current metal prices.

A significant fraction of the average annual loss of palladium to the domestic environment during processing, traditional uses, and disposal is in the form of innocuous palladium metal or palladium alloys. The concentration of palladium emissions as metal or oxide particulates or vapors to the atmosphere during domestic copper smelting is thought to be too low to pose any hazards. The 39,000 troy ounces of palladium expected to be lost annually in unrecycled electrical equipment is too widely disseminated to pose a health hazard. Losses of palladium metal or alloys used as petroleum-refining catalysts are estimated to be 50-340 troy ounces per year. Palladium metal or alloys lost to the environment from brazing alloys, dental alloys, jewelry, and glass coatings and from fabrication of palladium and its alloys (except in the case of electroplating wastes) are believed to be innocuous and of little environmental hazard.

However, a significant amount of the average annual loss of the estimated 160,000 troy ounces of palladium per year lost by the chemical industry must be in the form of palladium(II) compounds in wastewater streams. Moreover, loss of palladium(II) compounds by refiners and small electroplaters to wastewater streams may be sufficiently concentrated to be toxic to aquatic organisms.

Palladium and its compounds have not caused any recognized acute or chronic toxic effects in humans. However, the carcinogenicity of palladium(II) compounds in rats and mice, as well as the toxicity of these compounds to mammals, microflora, fish, and a higher plant (Kentucky bluegrass) are causes for environmental concern. Although palladium poses no urgent problem to humans, its toxicity to lower life forms suggests that palladium losses to the environment should be appraised regularly, that recycling of palladium-containing wastes should be increased, and that epidemiological studies in areas where palladium losses are expected to be high would be useful to determine whether palladium poses any human-health problems.

The appendices include the tables and figures and brief summaries of the physical and chemical properties and the analytical chemistry of palladium and its more common compounds.

I. INTRODUCTION

This document summarizes information available from published and company literature and industrial contacts on the natural occurrence, processing, uses, and disposal of palladium and its compounds. Suppliers and dealers were reluctant to disclose information on amounts sold to specific consumers. Amounts of palladium catalysts used in specific industrial processes could not be determined so that losses could only be estimated within a reasonable order of magnitude. The probable losses, distribution pathways, and ecological consequences of palladium presented in this report have been based primarily on subjective evaluations.

The findings of this study are presented in the following eight major sections: Sources, Economics, Processing, Uses, Losses to the Environment, Physiological Effects, Human Health Hazards, and Conclusions and Recommendations.

Appended to this report are tables supporting the textual discussions (Appendix A); tabulations of the physical properties of palladium, its common compounds, and alloys (Appendix B); a summary of the inorganic and organic chemistry of palladium compounds, including tabulations of the corrosion resistance of palladium to various inorganic compounds (Appendix C); and a discussion of analytic methods for determining palladium (Appendix D).

II. SOURCES OF PALLADIUM

The palladium content of the earth's crust is estimated to be 0.01-0.02 ppm.[1] Concentrations of palladium found in various natural minerals, soils and sediments, and other materials collected from a variety of geographical locations throughout the world are shown in Table A-I (page 61).

Palladium recovered for commercial use was obtained almost entirely from placer deposits until the 1920's. Since then the situation has been drastically altered by the discovery of plantiniferous lodes in Canada and South Africa. Today, a large fraction of the free-world production of platinum metals is recovered as a by-product of copper and nickel sulfide-ore refining.

Following are brief descriptions of the commercial sources and chemical forms of palladium found in nature and of the scrap materials that supply a large fraction of palladium consumed each year in the United States.

A. Natural Sources

The six platinum-group metals (platinum, palladium, ruthenium, rhodium, iridium and osmium) commonly occur in nature as two intergrown alloys.[2] However, these metals also occur as minerals in sulfide-ore bodies.

Alloys of the platinum metals are recovered primarily from placer deposits. The alluvial deposits in Canada were formed by weathering and erosion of dunite and serpentinite. Deposits of similar peridotites and perknites are the source of platinum metals in most placer deposits known in the world. In placer deposits, the two alloys are invariably mixed. One alloy has a high platinum content; a lower iridium content; and small contents of ruthenium, palladium, rhodium, and osmium. The second alloy consists mainly of iridium and osmium with considerable ruthenium; less rhodium and platinum; and very small amounts, if any, of palladium. Natural alloys and intermetallic compounds of palladium as well as the palladium content of these materials are shown in Table A-II (page 62).

The other workable deposits of platinum metals occur mainly as platinum minerals in nickel-copper, copper, and copper-cobalt sulfides that are genetically related to basic or ultrabasic rocks. The platinum metals occur mainly in sperrylite, cooperite, and other platinum and palladium minerals. Table A-III, page 63, contains a list of natural palladium compounds that have been identified. Small amounts of native platinum metals and alloys are often present in these mineral deposits.

Other natural sources include platinum minerals or native platinum alloys in copper and related ores indigenous to contact metamorphic and other types of ore bodies, including vein systems; native platinum metals in gold ores of quartz veins and other free gold ores; and meteorites. Noble metal abundances in meteoritic material are several orders of magnitude greater than in terrestrial material. The average crustal abundance of palladium is 0.01-0.02 ppm. Accretion of interplanetary dust by the earth may be the major source of palladium in deep-sea manganese nodules, rather than terrestrial-rock weathering.[3] Palladium was not detected in the principal U.S. rivers in a 1958-1959 study.[4]

Palladium has been detected in other minerals used commercially.. Less than 1-ppm palladium was found in coal taken from single mines in the Pittsburgh and Upper Freeport seams in Allegheny County, Pennsylvania.[5] Palladium has also been detected in 16 samples of calcium phosphate fertilizer, including three from North America.[6]

The principal sources of platinum metals in the world are the Bushveld Complex of South Africa, the Sudbury District of Canada, and the regions of Norilsk and the Kola Peninsula of the USSR.[7] Minor sources are placer deposits in Alaska, Colombia, Ethiopia, Japan, Australia, and Sierra Leone.

The Sudbury District of Ontario, Canada, contributes about one-third of the free-world production of platinum metals. The platinum-bearing ores are nickel-copper, copper, and copper-cobalt sulfides; native platinum metals or their alloys are usually absent in these ores.

South Africa produces most of the remaining two-thirds of the free-world's platinum metals, primarily from the nickel-copper lodes of Transvaal.

The estimated amounts of palladium in the platinum-group metals recovered from Canadian, Soviet, and South African sources are 42.9%, 60.0%, and 25.1%, respectively.8/ This compares with platinum contents of 43.4%, 30.0%, and 71.2% for the same sources. Platinum metals produced in the USSR are high in palladium and may account for as much as 80% of the world production of that metal.9/

Platinum metals have been found in 22 of the United States, but only Alaska is a major producer. The principal sources in Alaska are the Goodnews Bay District, Western Alaska, and a copper lode on Kasaan Peninsula in Southeastern Alaska. The latter was worked for several years, but is now closed. The Goodnews Bay District produced 3% of the world output of platinum-group metals in 1963.9/ Alaskan placer platinum metals, however, contain < 1% palladium.2/ In California and Oregon, small amounts of platinum metals are recovered as a by-product of gold placer mining.

Most of the platinum metals recovered from primary domestic sources are by-products of copper sulfide, lead, and gold ore refining.10/ Industrial contacts estimated the amount of palladium recovered from copper refining to be 75-95%11/ and 90%.12/ A small amount is recovered as a by-product of lead refining,13/ possibly as much as 15%.11/

B. Metal Scrap

Secondary recovery from metal scrap is the major domestic source of platinum-group metals, and accounts for about 10% of the new metal requirement. Most spent catalysts, however, are refined on a toll basis without transfer of ownership to the refiner.7/ "Toll refining," therefore, is categorized separately from secondary refining of metal scrap in that it does not represent a source of additional metal.

The three categories of scrap are: (1) mechanical fabrications of platinum-group alloys and spent solutions having a high precious-metal content; (2) filter media containing trapped precious metal and spent supported catalysts that can be refined but require prior preparation; and (3) materials with such small amounts of precious metal that smelting is the only economical method of removal, e.g., process residuals, mint and jewelers' sweeps, and composite materials containing both precious metals and commercially interesting base metals. Scrap sources include brazing alloys;

catalysts; dental scrap; thermocouple wire; telephone hardware; and electrical switches, relays, timers, and regulators.[14]

Electronics scrap is the major scrap source of palladium. Much palladium in electronics scrap is in the form of films on subminiature printed circuits on ceramic.[12] Integrated-circuit-type scrap is processed primarily for recovery of platinum-group and other precious metals.[15]

III. ECONOMICS

The production, distribution, and consumption of palladium as they relate to the U.S. economy are discussed in the two following subsections. The first reviews U.S. imports and exports, production, stocks, and sales of palladium and the difficulties encountered in attempting to reconcile differences in data reported for these activities. Summaries of these statistics are given in Tables A-IV through A-X (pages 64 through 70). The second subsection reviews the flow of palladium from world sources into the United States as well as the distribution and marketing of domestically produced palladium.

A. U.S. Supply and Demand

1. Imports: Platinum-group metal imports include refined metals, unrefined materials, crude ores and concentrates, grains, nuggets, and residues. The platinum-metal content of imported unrefined materials is not reported.[8] Total amounts of refined palladium imported into the U.S. in the years 1963-1971, however, are given in Table A-IV (page 64). Table A-V (page 65) compares these import totals for 1968-1971 with data compiled on imports of unwrought or partly worked palladium or palladium bars, plates, and sheets.

2. U.S. production: In 1968, domestic mine production of platinum-group metals supplied only approximately 10% of U.S. consumption and accounted for approximately 0.4% of world production.[8] U.S. palladium production is 3% of world output, although the U.S. consumes approximately 60% of world output.[16] Table A-VI (page 66) gives the annual U.S. refinery production of new and used palladium for 1961-1971.

About 90% of new palladium production is from domestic sources. Most of the new platinum-group metals produced in the U.S. are obtained as by-products of copper and gold refining. It is estimated that 75-95% of new platinum-group metals is derived from copper refining.

3. U.S. stocks: In the period 1961-1971, the General Services Administration Stock of palladium (Table A-VII, page 67) showed annual increases in 7 years. These increases represent purchases from imported or

domestic palladium and, thus, should be included when determining the U.S. demand of palladium in those years. In 1963, sales of palladium from the government stockpile would be considered a source of supply of palladium. Similarly, changes in the stocks of refiners, importers, and dealers and of the New York Mercantile Exchange (Table A-VIII, page 68) would be considered as either supply or demand figures according to whether they were negative or positive, respectively.

4. U.S. exports: The amounts of palladium that were exported annually in the period 1961-1969 could not be determined. Only data on U.S. exports of all platinum-group metals except platinum were reported (Table A-IX, page 69).

5. U.S. industrial sales: The annual U.S. palladium sales to industry are given in Table A-X (page 70). The average annual sales of palladium by industry in the period 1962-1971 were: chemical 25.74%, electrical 57.18%, dental and medical 7.95%, jewelry and decorative 2.81%, petroleum 2.49%, and glass 0.41%, with miscellaneous-and-undistributed figures accounting for 3.42%.

Actual sales may exceed the reported U.S. consumption because the data do not distinguish between shipments to firms for domestic use and shipments subsequently sent abroad. Changes in consumer stocks are unknown. Export data are not reported for palladium alone. Import figures are "imports for consumption" rather than general imports, a figure that should include material imported for processing and re-export. Re-exports are not counted in imports for consumption or domestic production and should not be deducted when attempting to calculate domestic consumption C. The obvious formula for such a calculation is C = Imports + Domestic Production-Exports-Δ Government Stocks-Δ Other Stocks. If total exports are deducted, consumption will be underestimated; yet the consumption estimate exceeds reported sales in about one-third of the years between 1950 and 1966, suggesting that the data are not wholly reliable. Therefore, reported sales have been assumed to closely approximate domestic consumption.[7]

B. Worldwide Sources for U.S. Consumption

The United States demand for palladium is met by the marketing of a wide variety of partly worked and finished items: sheet; wire, foil; ribbon, single crystals; semifinished jewelry; alloys; catalysts (gauzes, black, oxides, and supported forms); plating solutions; and various compounds.[1] The major part of this demand is supplied by importation of palladium from other countries and by domestic refining of new metal (see Table A-XI, page 71). The distribution of platinum-group metal from Canada, South Africa, the USSR, and the United States as they pertain to domestic palladium supply are summarized below.

Canada supplies about one-third of the free-world output of platinum-group metals. Some 90% of the Canadian output is produced by International Nickel Company of Canada, Ltd. (INCO), from nickel-copper lodes of the Sudbury District. Crude residues containing the platinum-group metals are produced from the Canadian ore and refined at Copper Cliff[2] or shipped to INCO's Platinum Metals Refinery in Acton, London, England, where the platinum metals are separated and recovered in pure form. The metals are eventually distributed in the U.S. by Engelhard Industries, Inc., Newark, New Jersey.

Falconbridge Nickel Mines, Ltd., produces the other 10% of Sudbury platinum-group metals. The nickel-copper mattes from its smelter at Falconbridge are refined at Kristiansand, Norway. The slimes from this process are refined and the platinum metals are recovered by Engelhard Industries.[2]

Two South African palladium producers (Impala and Atok Investments) sell their metals in the United States through distribution agencies. Impala sells their platinum-group metals through Ayreton Metals, a subsidiary of Western Platinum, Ltd. (owned by Lonrho, Ltd., London; Falconbridge Nickel Mines, Toronto; and Superior Oil Company, Houston). In 1970, it was reported that Impala planned to open two mines in the Merensky Reef; to send the matte to Falconbridge's refinery at Kristiansand, Norway, for recovering nickel and copper; and to refine the residue bearing the precious metals in South Africa.[18]

Platinum metals mined in the Lydenburg area of South Africa by Atok Investments, (Pty) Ltd. (owned by Anglo-Transvaal Consolidated Investment Company, Ltd.; U.S. Steel Corporation; and Middle Witwatersrand, Ltd.) are sold through the agent Leonard J. Buck, Inc., New York.

Another South African company, Rustenburg Platinum Mines, Ltd., produces platinum-group metals from osmiridium, which is mined chiefly for platinum, and from nickel-copper ores. The bulk of the Rustenburg production goes to Johnson, Matthey and Company, Ltd., in Great Britain for processing. Much of this metal is eventually sold in the United States by subsidiary companies, Matthey Bishop, Inc., and Johnson, Matthey and Company, Inc. Metal refined by Platinum Prospecting Company in South Africa is distributed by Engelhard Industries of New Jersey.

Soviet platinum metals are refined in the USSR and distributed to the U.S. directly (49% of 1966 U.S. palladium consumption) or via Switzerland, Germany, and other European countries.[7] Platinum metals imported from Western European countries (other than Norway and England) and Japan are probably Russian in origin since these countries do not refine platinum metals.[7] Occasionally, the USSR exports unrefined metal. The USSR intermittently stops selling on the world market.[19]

Much of the new palladium produced in the contiguous United States is the by-product of other metal production. Platinum metals from copper smelters and the by-product of copper and gold ores in the form of platiniferous residues and slimes are mined and refined by American Metal Refining, Division of American Metal Climax, New Jersey; Kennecott Copper Corporation, New York; and other electrolytic copper refineries (American Smelting and Refining Company, New Jersey; International Smelting and Refining, Division of Anaconda, New Jersey; and Phelps Dodge Refining, Division of Phelps Dodge, New York).[7]

The two largest refineries of platinum-group metals in the United States are Engelhard Industries and Matthey Bishop, Inc. (owned by Johnson Matthey and Company, Ltd., of the United Kingdom). Engelhard and Johnson Matthey distribute 80-85% of all free-world platinum.[7]

These two companies, along with a few other organizations, also process small quantities of concentrates and ores to recover platinum-group metals. The placer-ores concentrate (~ 500 troy ounces native ore per year) mined by Goodnews Bay Mining Company, is refined by Matthey Bishop of Malvern, Pennsylvania.[7,20] Colombian platinum metals imported as grain nuggets are refined by Engelhard; Matthey Bishop, Inc.; and Goldsmith Brothers (a subsidiary of National Lead Company).[7] Engelhard Industries also refines concentrates and platinum-bearing matte from the Brakesprite Mine in the Republic of South Africa. The concentrate from placer deposits of gold-platinum ores is mined and refined by Yuba Development, a Division of Yuba Consolidated Industries, California.[7]

IV. PROCESSING OF PALLADIUM ORE, METAL, AND SCRAP

The mining, refining, reclamation, and production processes used for palladium are described in this section of the report. Losses from these operations as well as those resulting from the use and disposal of palladium-containing materials are discussed in Section VI (page 41). A tabulation of U.S. metal works that process palladium is given in Table A-XII (page 72).

A. Mining

The specific method of mining employed to recover palladium-containing ore is dependent on the ore body. Alluvial deposits at Goodnews Bay were recovered by dragline until 1957 and dredging since then.[21] The alluvial deposits being mined include the Goodnews deposit in Alaska, gold placers in the Ural Mountains of the USSR, and the Colombia deposits. Ores from these deposits are generally concentrated by gravity separation. This same technique is used in South Africa to concentrate osmiridium.

Copper and nickel ores that contain platinum metals are commonly mined by open-pit methods. The platinum-group metals in the Sudbury District of Ontario, Canada, are mixed with the sulfide ores of copper and nickel. Concentrates of the copper and nickel sulfides are obtained by magnetic and flotation techniques.

B. Refining

In the INCO recovery and separation process (Figure 1), the copper-nickel concentrate is roasted and then blown to give a Bessemer matte. The bulk of the platinum metals separates from the Bessemer matte during slow cooling and is removed magnetically. After smelting and magnetic separation, the enriched alloy, containing all six platinum-group metals, is electrolytically refined. The anode slimes produced during electrolysis contain the platinum metals. (Smaller amounts of platinum metals are obtained from nickel refining by the Mond carbonyl process and from electrolytic refining of nickel concentrates.[17/]) In conventional treatment for base-metal removal, the anode slimes are often sulfated at 600°F and then extracted with water to remove residual nickel and copper.[22/]

After extracting with water, the concentrate that remains contains the precious metals. To recover the precious metals, the concentrate is calcined in air at 1500-1800°F to eliminate sulfur, selenium, and arsenic impurities.

The concentrate from the anode slimes is sent by INCO to Acton, England, for refining at the Mond Platinum Metals Refinery.[23/] The nickel oxide sinter, obtained by roasting the nickel sulfide concentrate from the differential flotation process, is treated by the Mond carbonyl process to give nickel pellets. The process residue is worked up to give a platinum-metals concentrate for refining at Acton.

Palladium is separated from the platinum-metals concentrates as shown in Figure 2. The pure palladium salt $Pd(NH_3)_2Cl_2$ is ignited with hydrogen to give palladium sponge. The latter is reduced with hydrogen to remove oxide films and to give $> 99.9\%$ pure palladium.[23/]

Up to 0.25% of the platinum metals contained in the concentrates is lost. Less than 0.5% is temporarily retained in furnace slags, which are retreated at the nickel smelter. Losses occurring during extraction of copper and nickel are small. It is estimated that 90% of the precious metals contained in the original ore are recovered.[23/]

Platinum-bearing nickel ore from the Republic of South Africa is crushed, ground, and gravity concentrated to give a "metallics" concentrate that contains 22% platinum metals, about two-thirds of the precious metals in the ore. It is sent to Johnson Matthey Company, Ltd., at Brimsdown, England, for refining.[17/]

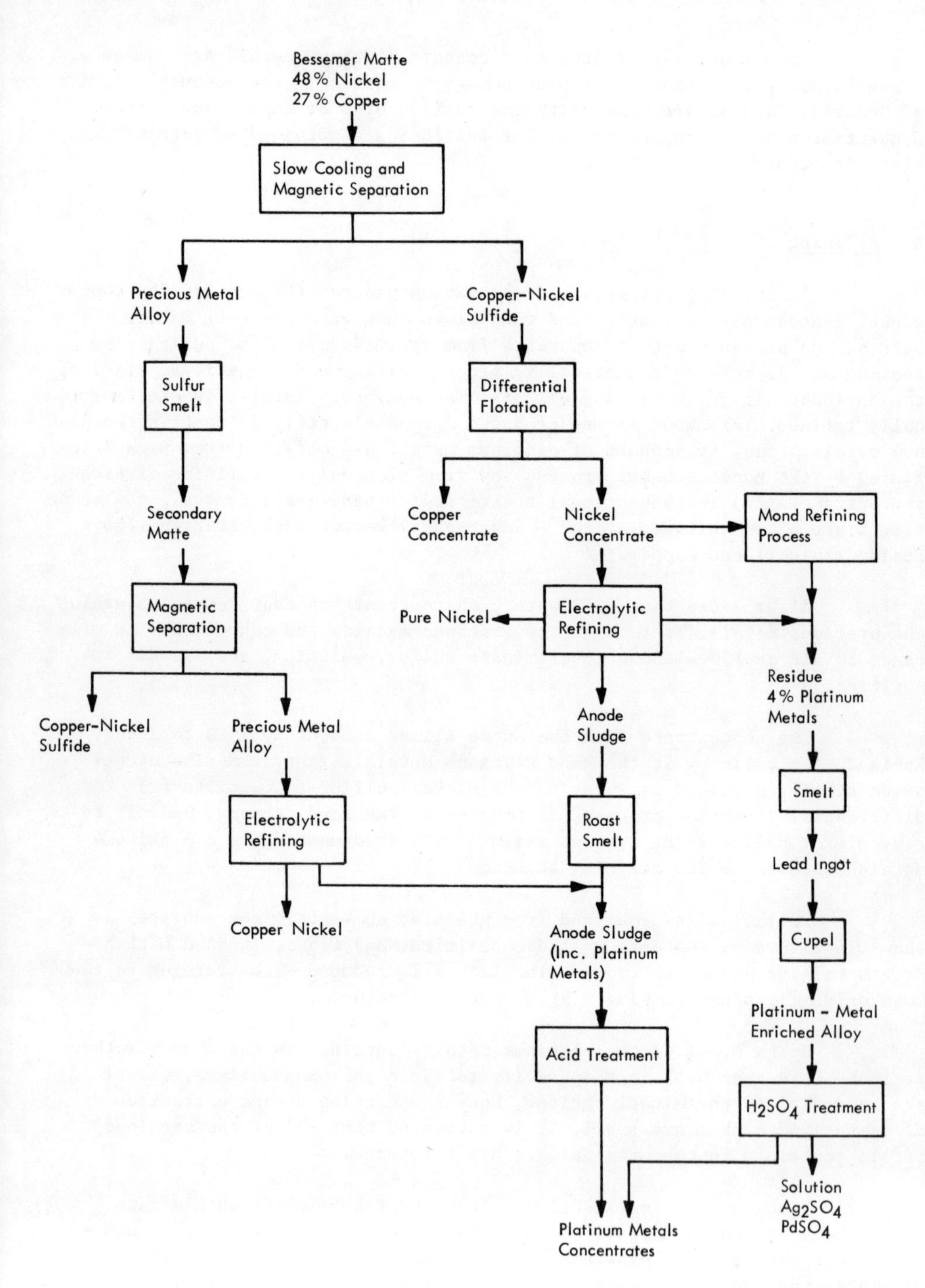

Figure 1 - Recovery of Platinum Metals from the Sudbury Ores[17/]

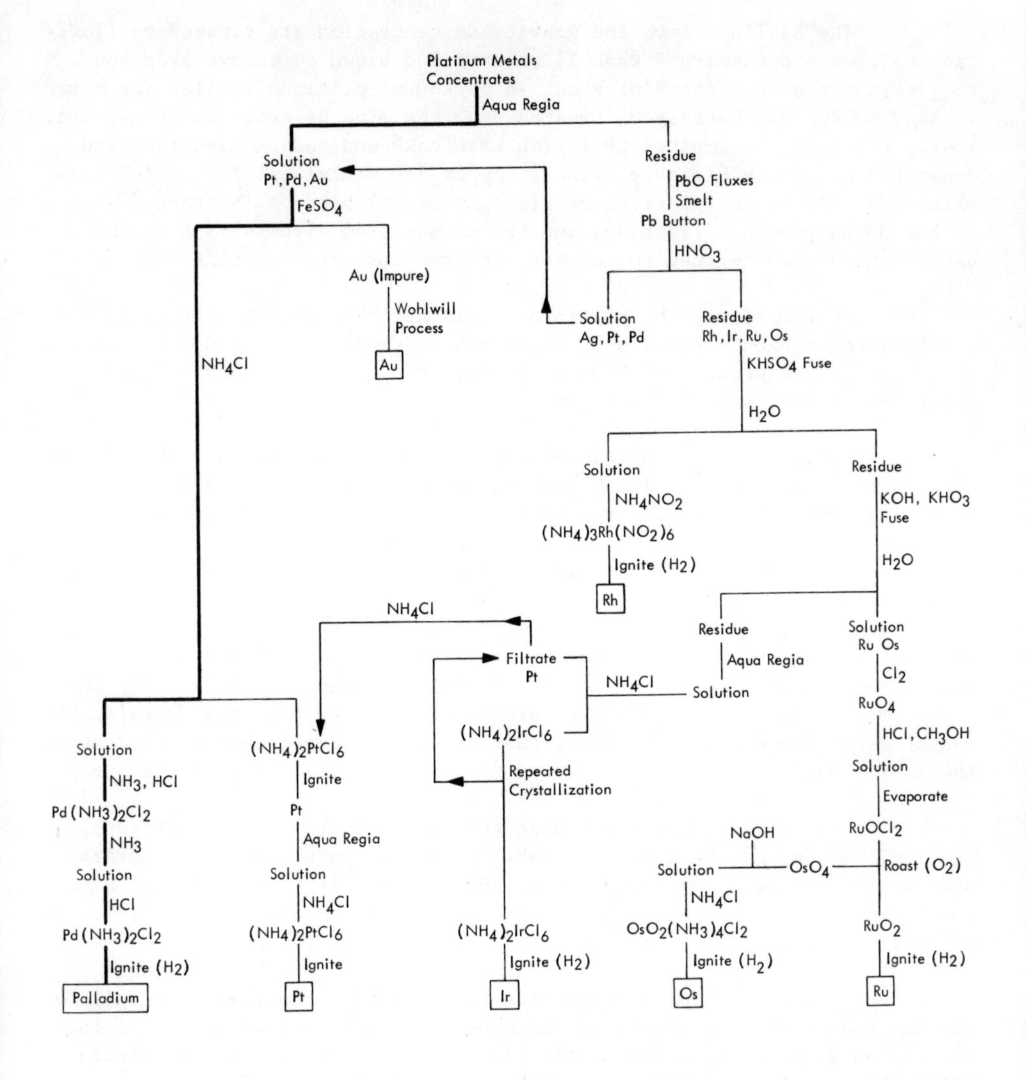

Figure 2 - Palladium Recovery Process[17]

The tailings from the gravity concentration are treated by flotation to give a concentrate that is smelted and blown to remove iron and to give a matte that contains about 48 oz/ton of platinum metals. The matte is sent to Johnson Matthey or treated near the mine by Matte Smelters, Ltd. There, the matte is smelted by fusion with coke and sodium bisulfate and separated to give ultimately blister copper and nickel sulfide, which contains the bulk of the platinum metals. The nickel sulfide is treated much as the Sudbury nickel fraction, and the slimes from electrolysis of the nickel anodes are refined to recover the precious metals.[17,24]

At Brimsdown, the platinum metals are extracted by enriching the platinum-containing materials to about 65% platinum metals and then treating the enriched product with acids to separate the individual platinum-group metals followed by final refining.

Mining of crude platinum in placer deposits furnishes only a small part of total production. The mining and processing techniques for recovering crude platinum from placers are similar to those used for recovering gold.

A major portion of the U.S. domestic production of platinum metals is recovered as a by-product of the copper industry.[10] It is estimated that 1 oz of the platinum metals is recovered from 35 tons of copper produced; however, no serious effort is made to recover the precious metals in high yield. No data were found on amounts of platinum-group metals lost to the environment from copper and nickel ores not processed for these metals. It can be generally assumed, however, that the platinum-group-metal content of these ores must be too low to justify the cost of additional processing.[8]

Most of the copper mined in the United States is from Arizona, Montana, Nevada, New Mexico, and Utah. There are seven smelters in Arizona and one each in Utah, Montana, Nevada, New Mexico, Tennessee, Texas, Michigan, New Jersey, and Washington.[25] Platiniferous residues are refined by copper refiners in Maryland, New Jersey, Texas, Utah, and Washington.[10]

The major steps in producing copper metal from low-grade sulfide ore are beneficiation, roasting, smelting, and refining. In the U.S., the ore is concentrated near the mines. Flotation is commonly used to concentrate the ore after it has been crushed, ground, and classified.[8]

Figure 3 depicts typical copper smelting and electrolytic refining practices. Copper sulfide ores and concentrates are roasted at 1200°F (649°C) to regulate the amount of sulfur for efficient melting and to remove volatile impurities such as antimony, arsenic, and bismuth. The concentrate from some sulfide ores requires no roasting prior to reverberatory smelting at 2400°F (1316°C). Roasted and unroasted materials are melted with fluxes in a reducing or nonoxidizing atmosphere to produce a copper matte, comprising a mixture of Cu_2S and FeS and containing 15-50% copper. The precious metals dissolve in this matte. Production of a high-grade matte can result in poor recovery of precious metals.

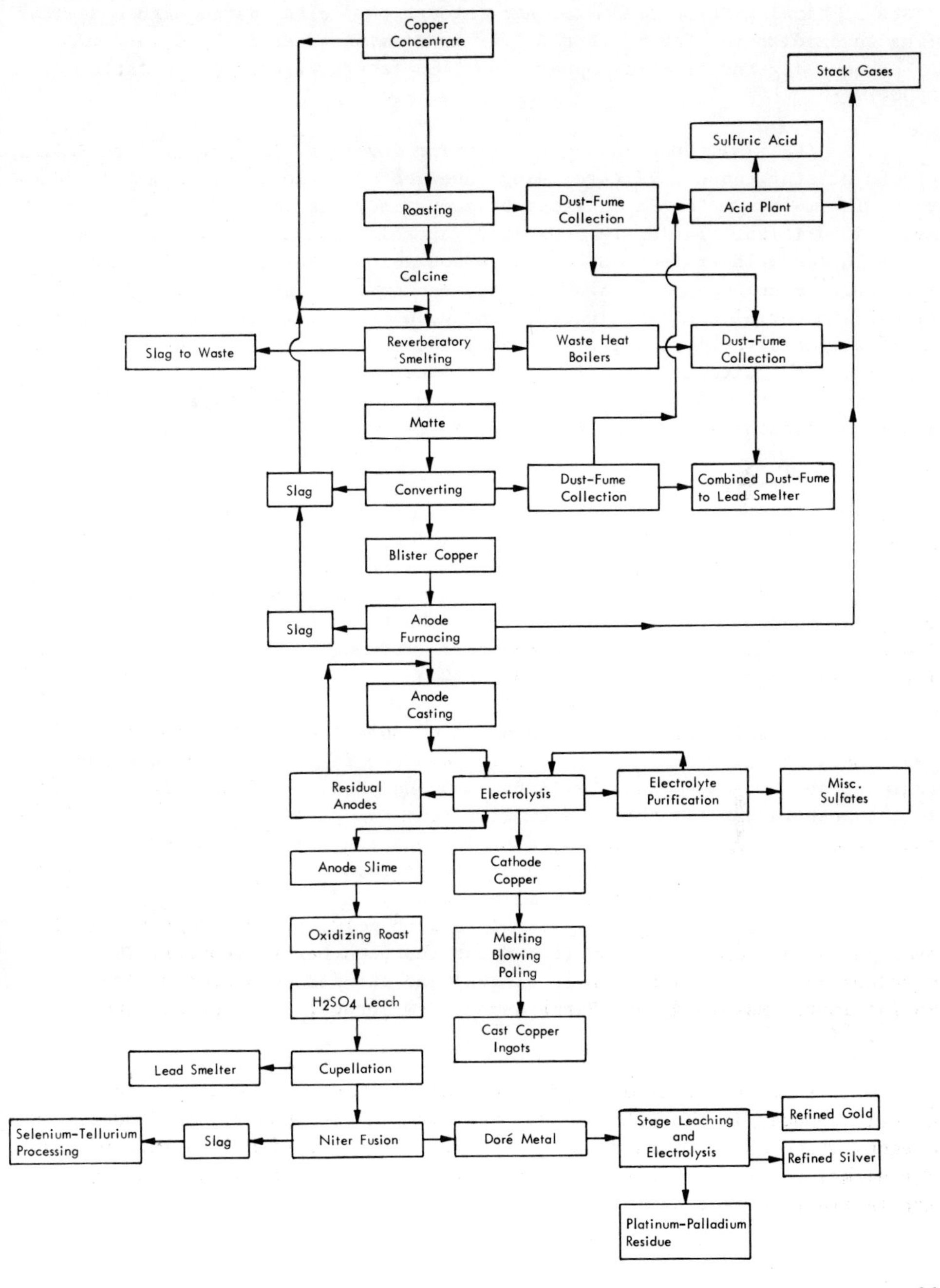

Figure 3 - Typical Copper Smelting and Electrolytic Refining Practice[26/]

The final stage in the smelting process is converting, which consists of blowing thin streams of air through the molten matte with a silica flux to oxidize the FeS so that sulfur is eliminated as SO_2 and to form a ferrous slag and blister copper. The temperature of the molten matte is 1800°F.[8]

The final step in copper refining involves electro- or fire-refining of the blister copper. Fire-refining involves oxidation, fluxing, and reduction. The molten metal is agitated by compressed air to oxidize sulfur and metal impurities. Fluxing removes lead, arsenic, and antimony. The copper oxide in the melt is reduced by green-wood poles inserted below the bath surface.[8] Precious metals are not oxidized in fire refining.[27] When the original material contains insufficient gold or silver to justify recovery or if a special-purpose silver-containing copper is desired, the fire-refined copper is cast directly into forms for industrial use. If the recovery of precious metals is warranted, fire-refining is carried only far enough to insure homogenous anodes for electrolytic refining.

Fire-refining capacity in the United States in 1968 was 365,000 tons, whereas 2.3 million tons of copper was electrolytically refined. Approximately 60% of U.S. electrolytic-refining capacity is in six Atlantic Coast refineries. Fire-refining is done in Michigan, New York, New Jersey, New Mexico, and Texas.[8] Of the 16 U.S. copper smelters, five produce only blister copper. Other fire-refine at least part of the blister copper.[27] Anode slimes remaining after electrolytic refining of copper are processed for platinum metals. (See Figures 3 and 4.)

Palladium is also a by-product of some lead production. Typical lead smelting and refining practice is depicted in Figure 4. As a specific example, at the Asarco Lead Refinery in Omaha, Nebraska (the third largest in the western world), the lead bullion from the smelters is oxidized, forming a slag of oxidized antimony, arsenic, and tin that is skimmed off. The softened bullion is treated with agents that combine with copper and rise to the surface for removal by skimming. Zinc is added to the molten lead bullion so that, on cooling, it combines with gold, silver, copper, platinum, palladium, and tellurium in a crust that can be skimmed off. The precious metals are cast as Doré (ingot) and shipped to Asarco's plants in Baltimore, Maryland, and Perth Amboy, New Jersey, to be parted and refined.[13]

Sulfide nickel ores often contain palladium. The only primary nickel producer in the U.S. in 1968 was Hanna Mining Company of Riddle, Oregon, which produces annually 13,000 short tons of nickel in ferronickel. The nickel occurs as lateritic nickel oxide.[28] Apparently no platinum metals are recovered from this ore.

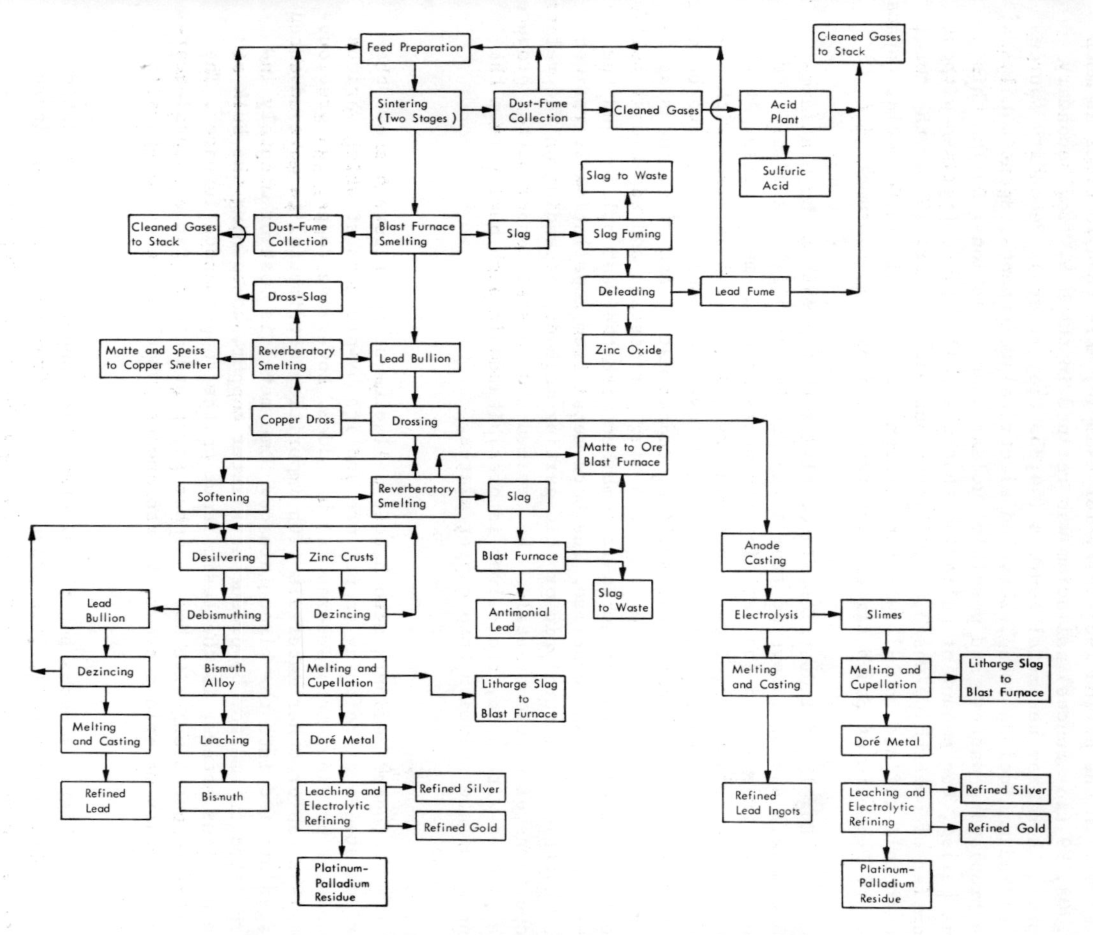

Figure 4 - Typical Lead Smelting and Refining Practice[26]

C. Secondary Recovery

Palladium recovered from secondary sources, i.e., from scrap material and from toll processing of spent catalysts, provides most of the metal used each year (see Table A-VI, page 66).

In the period 1967-1971, for example, more than 18 times as much (1,008,199 troy ounces) palladium was refined in the U.S. from secondary sources than from new metal sources (54,833 troy ounces). Secondary sources of palladium include scrap, recycled electrical components, spent catalysts, electroplating wastes, and possibly nuclear waste solutions. During this same period, the amount of palladium that was toll refined (refined without transfer of ownership) was 3,873,095 troy ounces. However, 16% of the total amount toll refined in 1971 was for foreign owners mainly in Colombia, Canada, and the Republic of South Africa. The rest is from U.S. consumers.29/

Engelhard Industries, Inc., claims that ≥ 95% of the platinum-group metals can be recovered from scrap.30/ Figure 5 gives a simplified flowsheet for smelting and refining secondary precious-metal scrap. At the precious-metal smelter, low-tenor secondary silver scrap is mixed with recycled metallic lead, copper pyrite, and fluxes and smelted in a reverberatory furnace to produce impure lead bullion, copper matte, and slag. The slag is resmelted in a blast furnace to give a slag for discard and additional bullion and matte. Clean metallic scrap is combined with the lead bullion and the mixture is melted and cupelled. The Doré metal from cupellation is refined by conventional electrolytic or acid parting methods, depending on the relative proportions of precious metals. The crushed and roasted copper mattes produced are leached with acid, and silver is precipitated in high purity from the copper sulfate solution onto copper plates.

Copper refiners also handle palladium scrap. Figure 6 shows the various stages of refining copper ore and its by-products at which various grades of precious-metals scrap can be added. Low-grade copper and precious-metals scrap or other materials with impurities detrimental to more advanced stages are fed into the blast furnace. Some metal is lost or minutely incorporated in smelter by-products. Blister copper, containing the bulk of the precious metals, is the basic input to the copper anode furnaces. The blister copper along with No. 2 copper scrap and other precious-metal-bearing secondary materials are cast into anodes, which are electrolytically refined.31/

Figure 7 shows how the electrolytic copper slimes are treated in somewhat more detail. This process differs from that shown in Figures 1 and 2, in which the precious metals from Sudbury ore copper slimes are separated almost entirely by chemical means. International Nickel, however, does furnace the treated slimes to give Doré metal.

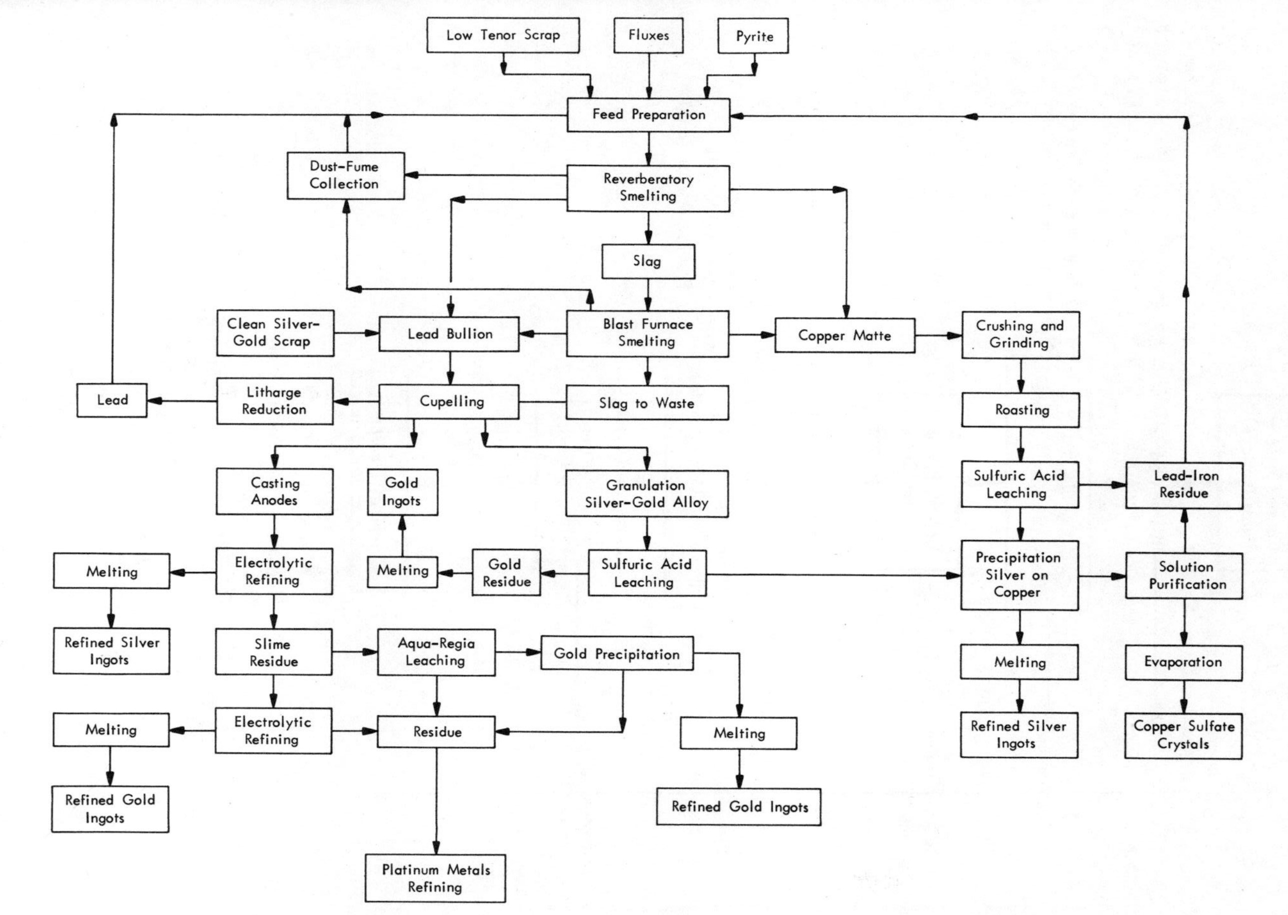

Figure 5 - Smelting and Refining of Secondary Precious-Metal Scrap26/

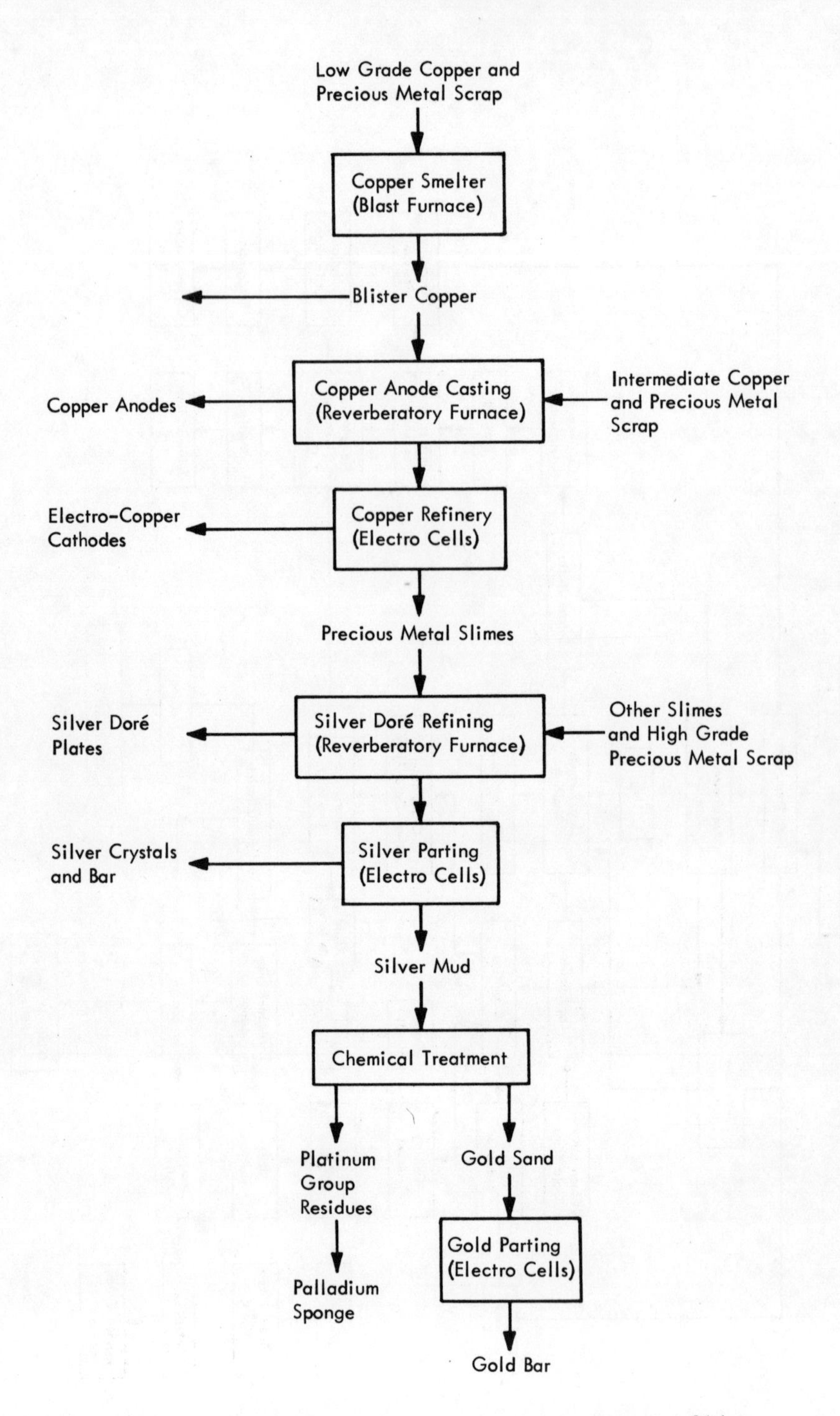

Figure 6 - Secondary Recovery by a Copper Refiner[31]

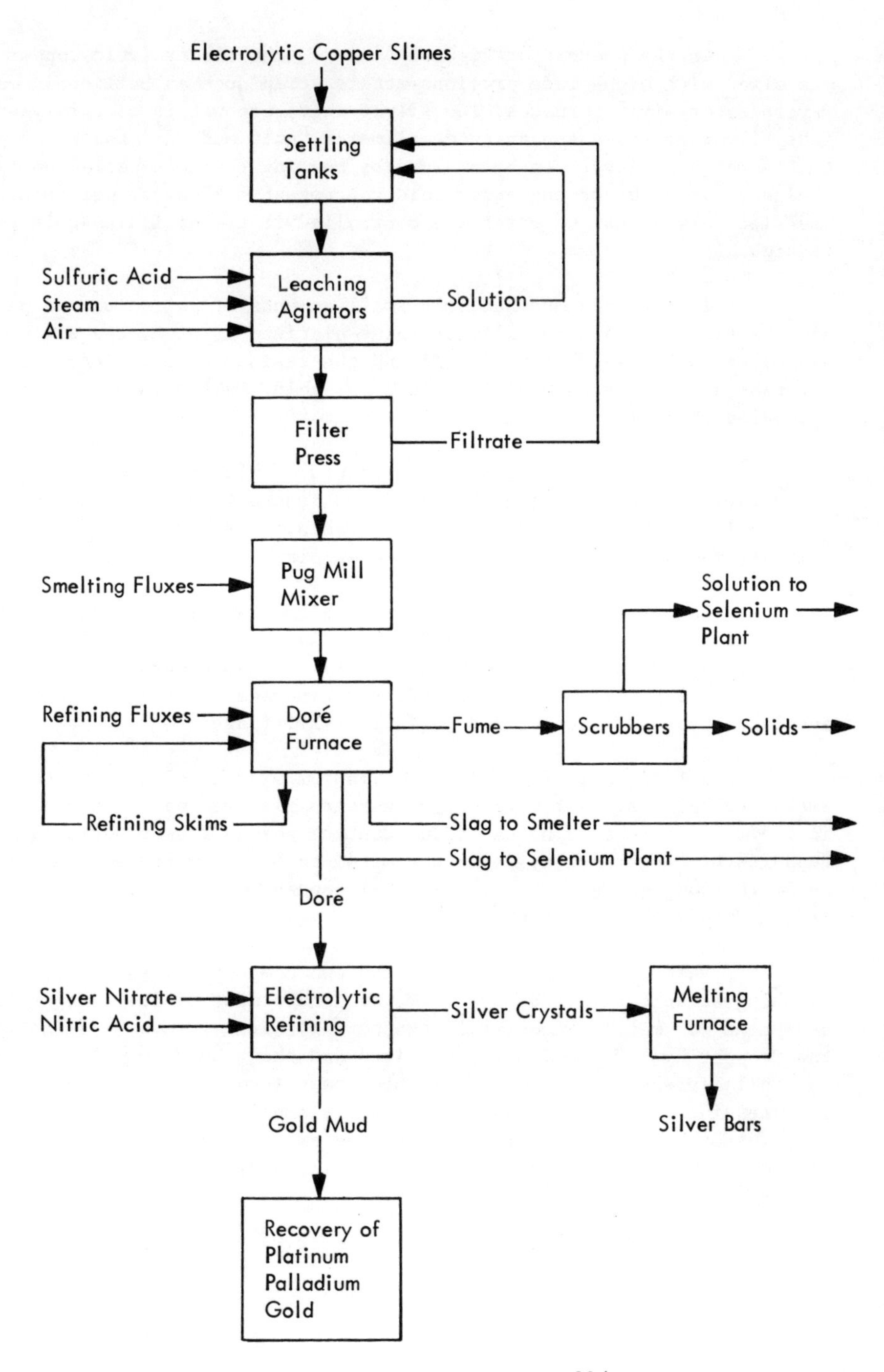

Figure 7 - Silver Refinery Flowsheet[32]

In the process of Figures 6 and 7, the electrolytic copper slimes are mixed with high-grade precious-metals scrap such as bullion in Doré reverberatory-type furnaces. The silver-anode product is electrolyzed to give silver crystals and an anode slime or "gold mud."[31] Losses of palladium to the refined silver can be avoided by keeping the gold-palladium ratio in the anode at ≥ 10 and the anode gold content at < 45 parts per thousand. The electrolyte must be carefully controlled if the palladium-gold ratio is high.[32]

If the gold mud has only small amounts of palladium and platinum, and the ratio of gold to palladium-plus-platinum is high, the silver is dissolved in boiling sulfuric acid and the residue is cast into anodes for electrolysis for the Wohlwill process for gold. Palladium and platinum are recovered chemically from the electrolyte.[32]

If the gold mud has large amounts of platinum and palladium and their ratio to gold is high, the gold mud is treated with aqua regia from which gold is precipitated followed by Wohlwill process purification. The filtrate is treated chemically to recover platinum and palladium.

Engelhard Industries rerefines high-grade scrap, which contains 30-40% precious metals. American Smelting and Refining and AMAX refine low-grade scrap. Engelhard also refines easily separable combinations containing less palladium.[33] Other secondary refiners as well as primary refiners and fabricators are identified in Table A-XII (page 72).

Only a small fraction of palladium used in electrical components, except telephones, is recovered.[34] Western Electric was issued a U.S. patent in 1957[35] for recovering palladium contact points from brass springs and supports by electrolysis, using the metal to be recovered as the anode. At pH 5-7.5, 600 amp, and 15 V, the nickel bonding metal dissolves and palladium drops to the cell bottom.

Western Electric operates its own company, Nassau Smelting and Refining at Staten Island, New York, to handle most of its scrap telephone equipment for metals recovery.[15] Western Electric recovers palladium through the "copper cycle" previously described and shown in Figure 6, in which the palladium-bearing copper is melted, cast into bars, and treated in anode furnaces for copper refining; the copper slimes are refined for silver in Doré furnaces; and the silver cell slimes are treated to recover their palladium content. According to Mr. Howard Martin of Martin Metals, there are no losses of palladium.[12]

Other electronic scrap containing palladium includes aluminum-cased components from aircraft radar, communications, tracking and warning systems, and industrial computers and controls. The electronic components have copper pins, contacts, and connectors that are plated with silver, gold, and other precious metals to insure better electrical contact and corrosion resistance. The aluminum ingots derived from this sweated electronic scrap can be processed by electrolytic methods to recover aluminum and concentrate the base and precious metals in an enriched anode. Such ingots contain most of the aluminum and part of the precious metals and copper in the original scrap (e.g., ingots from military scrap contained 63.90% aluminum, 22.23% copper, 128 ounces silver per ton, and 1.05 ounce palladium per ton).[36/]

Engelhard and Johnson Matthey are probably the two largest refiners of spent hydrocracking catalysts.[12/] Matthey Bishop claims that 98% of all platinum group metals used in the petroleum industry is reclaimed.[37/] Palladium can be recovered from the spent hydrocracking catalyst by the secondary refiner in 90-95% yield, depending on how badly the catalyst has been contaminated during use.[33/] Matthey Bishop recovers palladium from the catalyst by chemical treatment.[12/]

Union Oil Company may regenerate palladium zeolite molecular-sieve hydrocracking catalysts that have been deactivated by water by alternate hydrogenation and oxygenation treatments at 700-950°F [U.S. Patents 2,197,397 and 3,197,399, (1965)].[38,39/]

One of the large uses of palladium is as a catalyst in the production of organic chemicals. Following are brief descriptions of methods that have been identified to recover palladium catalysts in certain uses.

Palladium can be recovered from spent catalysts (palladium and cupric chlorides) that have been used to produce carbonyl compounds by extracting with acid and depositing on an inert surface ≥ 125°,[40/] simultaneously oxidizing during the acid extraction and precipitating as K_2PdCl_4[41,42/] or simultaneously oxidizing and then cementing with a less noble metal,[43/] or regenerating the reduced palladium with hydrochloric acid and oxygen.[44/] Atlantic Richfield's process[41/] claims to give quantitative recovery.

Palladium compounds, such as palladium chloride and tetrachloropalladates(II), can be economically recovered from aqueous solution by treating with olefinic complexing agents, isolating the complex, and thermally decomposing the complex [Eastman Kodak Company, Ger. Offen., 1,816,162 (1969)].[45/] Palladium ions can be selectively complexed by and eluted from polyamidoxime fabrics in the presence of other metals such as plutonium, gold, platinum, rhodium, iron, thallium, vanadium, uranium, ruthenium, copper, nickel, and chromium by controlling the pH [NOPCO Chemical Company, U.S. 3,088,799 (1963)].[46/]

Spent platinum-metal hydrogenation catalysts are usually regenerated by oxidation, steam, heat, and (or) solvent extraction.[47]

Deactivated catalyst from the anthraquinone process for preparing hydrogen peroxide is reactivated by treating with steam containing about 10% water at 80-200° [FMC Corporation, U.S. 3,112,278 (1963)].[47]

Reduced alumina-supported palladium catalysts used for producing hydrogen peroxide are recovered by dissolving with 8 N hydrochloric acid and chlorine, precipitating on an aluminum sheet, and redissolving in hydrochloric acid and chlorine to prepare fresh catalysts [Laporte Chemicals, Ltd., Britain, 922,021 (1963)].[48] Palladium can also be recovered from spent palladium-on-alumina catalysts by volatilizing in a stream of aluminum chloride at 200-1600°F [Universal Oil Products, U.S. 2,828,200 (1958)].[49]

Spent palladium-on-carbon catalyst used in hydrogenating succinic anhydride to butyrolactone is regenerated by treating with dilute sodium hydroxide, water washing, and exposing to air [FMC Corporation, U.S. 3,214,385 (1965)].[50]

Palladium surface deposits, e.g., on distillation columns, can be treated at 10-150°C with an aqueous oxidizing solution containing 0.5-2.0-moles cupric chloride per liter solution at ≥ 10% volume percent acetic acid and an absolute emf 500-750 mV to give a solution useful as a component of an olefin-oxidation catalyst [Eastman Kodak Company, Ger. Offen., 1,926,042 (1969)].[51]

Farbwerke Hoechst A.-G. holds a 1965 patent[52] for recovering palladium from acidic solutions by reducing to palladium metal with hydrogen and extracting with an oxidizing acid or an oxidizing agent in the presence of a nonoxidizing acid.

PPG Industries, Inc., holds a 1970 patent[53] for recovering traces of metal ions, including palladium, such as are found in sewage or seawater, by electrolysis with deposition on a porous graphite or carbon-black cathode.

Electroplaters frequently recover the palladium content of spent electroplating baths by precipitating with organic reducing agents such as hydrazine or formaldehyde, unless the bath is contaminated.[12,54] The metal content of a contaminated bath is precipitated as a sludge and sent to a secondary refinery.[11] The secondary refinery commonly recovers palladium from plating baths by ion exchange or precipitation by some metal higher in the electromotive series.[12]

One secondary refinery, Martin Metals of Los Angeles, California, recovers 99-100% of the palladium from spent plating baths. The value of palladium in such baths may be 30¢/gal. Baths containing only 2-15¢ worth of palladium are discarded to the sewer.[12] At $84/troy ounce of palladium, this would be 1.97-14.8-ppm palladium.

Typical effluents from plating ABS-type plastics with nickel, copper, and chromium are reported to contain 0.03-0.05 mg/liter palladium. The palladium is probably used to sensitize the plastic.[55]

Palladium, rhodium, and technetium can be recovered from nuclear waste solutions (from the fission of ^{235}U and ^{239}Pu) by ion exchange.[56,57] Power reactor fuels exposed to 25,000 MWD/ton should have a Pd concentration of 880-1,660 g/metric ton with an activity of 100 μCi/gram.[58] Spent fuel fission products are expected to yield 96,400 ounces of palladium per year by 1982[59] and to yield amounts equal to those from mineral sources after 1990.[58]

Palladium in supernatant solutions of the high-salt wastes stored following the recovery of actinides from irradiated uranium was recovered in 97% yield by loading on a packed column of Aliquat-336 [tricapryl(monomethyl)-ammonium chloride] on an inert support followed by three cycles of elution with nitric acid and ammonium hydroxide.[60]

Techniques that were evaluated by the U.S. AEC for decontaminating precious metals[61] include surface treatment with acidic solvents and fused salts and complete dissolution followed by purification, best done by solvent extraction and precipitation.

It is doubtful that the palladium and rhodium theoretically available from nuclear reactors will be economically recovered and utilized because of their radioactivity and the problems associated with their extraction from highly active fission products.[62] Fission-product palladium contains about 15% ^{107}Pd, a stable beta emitter of half life 7,000,000 years. The beta radiation energy is 0.035 MeV. Such material might cause skin damage if used in jewelry. Silver reclaimed from scrap silver-palladium alloys made with this reactor-fuel by-product could not be used for photographic film unless all the palladium was removed. Since most used palladium is recycled many times, the entire pool of palladium could eventually become contaminated by radioactivity. The use of fission-product palladium is not likely to occur, certainly not so long as reserves and mining development are adequate.[63]

D. Metal Processing

The physical properties of palladium make it well suited for use as an alloy and as a surface coating. The following briefly describes processes by which palladium alloys and coatings are produced.

1. Palladium metal and alloys: Palladium, like gold, silver, and platinum, is soft, malleable, and ductile. It is easily formed by conventional methods and is usually hot forged or rolled at 1200-1400°C from the ingot to intermediate sizes and thereafter cold worked.[64,65] Small ingots are often reduced to the finished size by cold working. Impurities and the degree of cold working alter the mechanical properties of palladium. Although palladium work hardend somewhat more rapidly than platinum, it can withstand drastic working and forming operations and can be fabricated into leaf 0.000004 in. thick.[66]

The hardness of wrought noble metals in the hard-worked condition is higher than that in the annealed state. Electrodeposits of palladium are very much harder.[67] The microhardness for electroplated palladium ranges from approximately 50 to 450 kg/mm^2.[68]

Palladium can also be handled by melting and casting with deoxidation prior to pouring, e.g., by calcium, calcium boride or aluminum.

Alloys of palladium to be used for resistors, brazing alloys, thermocouples, purification, jewelry, process equipment, and various other uses such as in prosthetic dentistry and orthodontia are designed to exhibit specific properties such as high-melting points, corrosion resistance, or hardness. Elements such as rhodium, ruthenium, and nickel increase the hardness of palladium alloys. Cold work and addition of elements such as ruthenium and rhodium increase the tensile strength and limit of proportionality.[17] Palladium alloys are also relatively easy to fabricate by conventional techniques.

Precautions must be taken during fabrication to minimize gas absorption during melting and to prevent surface oxidation of palladium during annealing. Palladium may be annealed in air at 850°C, but it must be quenched in water from above that temperature to avoid a surface palladium oxide tarnish. Palladium and its alloys are annealed under atmospheres of nitrogen, 95% nitrogen-5% hydrogen mixture, argon, or steam or *in vacuo* at temperatures up to 1100°C to avoid surface oxidation. Tarnish films can be removed by pickling with 10% solutions of hot sulfuric acid or tartaric acid.[64,66,69]

Small lots of palladium alloys used for such purposes as jewelry and dental applications are torch melted. Dental alloys can be melted with a gas-air torch. Inductive heating is best for producing large palladium melts.[69]

Palladium is not easy to solder.[70] An oxidizing oxyacetylene flame is used to solder palladium with platinum solders melting at 1100-1300°C. To avoid surface and subsurface damage to palladium jewelry during hard soldering at ≤ 1400°C, the oxyacetylene flame is operated slightly oxidizing. Palladium jewelry can be soldered by a gas-air torch and lower-melting white-gold solders. Jewelry made of 95.5% Pd-4.5% Ru alloy can also be soldered electrically. Solders for dental gold alloys do not contain palladium.[69]

Brazing methods with palladium alloys include torch brazing, furnace brazing, induction brazing, salt-bath brazing, dip-brazing, and resistance brazing. Furnace brazing with a reducing atmosphere is the preferred method for joining copper components.[71]

2. Palladium coatings: Palladium is used in many coating applications because of its noble-metal characteristics. It is often most economical to use only a thin coating of the metal in such instances. Palladium coatings can be produced by a number of processes including electroplating, electroless deposition, thermal decomposition of organic compounds, vacuum deposition, sputtering, and cladding.

a. Electrodeposition: A wide variety of baths are used in the electrodeposition of palladium ranging from the palladous chloride high-palladium-ion bath for electroforming applications to the low-ion baths of the tetranitrate. tetraammine, and more complex types.[72] Typical plating solutions and conditions are discussed by Wise.[73]

Baths containing the tetraammine complex made by dissolving $Pd(NH_3)_2Cl_2$ in ammonia were used for years at INCO's Port Colbourne, Ontario, refinery as a final step in refining palladium. Modifications of that bath are used for applying decorative deposits.[73]

Although thin (< 0.1 mil [2.5 μ]) palladium deposits were used for years on jewelry and electrical contacts, thick (> 2 mil [50 μ]), deposits often had high internal stress, which led to cracking, or had low adhesion.[74]

Before 1964, the only production baths in general use were based on palladium diammino nitrite $Pd(NH_3)_2(NO_2)_2$ [palladium "P" salt] or palladium tetraammino nitrate $Pd(NH_3)_4(NO_3)_2$. Organic and other impurities caused excessive stress and cracking and resulted in unattractive deposits.[75] Electrolyte solutions were similar to those used in plating platinum but required lower applied voltage and more careful pH control.[17] Baths were developed based on a sulfamate solution that gave more attractive deposits but were more sensitive to contamination, and the stressed deposits were not crack resistant.[75]

Nonproprietary aqueous electrolytes were developed at the Bureau of Mines in 1969 from which palladium can be electrodeposited in thick, adherent, and coherent layers. Palladium was deposited from baths containing palladium and ammonium chlorides and hydrochloric acid (with expendable palladium electrodes) or $Pd(NH_3)_2(NO_2)_2$ and sulfamic acid (with insoluble electrodes). The former electrolyte was the most satisfactory. Palladium deposits ≤ 20.2 mil thick were formed on molybdenum, niobium, and tungsten following a cathodic pretreatment in a fused cyanide electrolyte containing platinum-group metal salts. Coatings ≤ 16.8 mil and ≤ 24 mil thick were formed on nickel and titanium, respectively, after etching with aqueous acid or salt solutions.[74]

Palladium plating baths deteriorate with age. Only new nitrate and nitrite solutions give ductile, pore-free deposits. American plating solutions that operate cold without free ammonia give microcracked deposits, which are useless for corrosion protection. Solutions with a high ammonia content, which operate at near-boiling conditions, present control problems.[70]

Palladium may be coated electrolytically on electronic components such as end connectors of printed-circuit boards, reed switches, and spring contacts.[76] Since the methods of electrodepositing palladium have improved, palladium plating may be used as an alternative to gold in printed circuits at one-third the cost of gold.[77,78]

Electrodeposition from aqueous solutions has been used to produce binary alloys of palladium with platinum, rhodium, and ruthenium.[79]

b. Electroless deposition: An electroless plating solution may contain, for example, palladous ion, as e.g., $Pd(NH_3)_2Cl_2$, hydrazine, ammonia, and EDTA; and give 0.006-cm-thick deposits on transition metals. Electroless plating with hydrazine as the reducer is not suitable for direct palladium deposition on copper or its alloys, zinc, or magnesium but is especially suitable for palladium-coating small nickel parts by barrel-plating. Palladium has been deposited electrolessly on silverware and decorative items.[73]

Palladium or palladium alloys containing 6% nickel, 10% cobalt, or 36% zinc can be deposited electrolessly on steel, copper, nickel-plated steel, brass, gold, glass, or plastics from stable plating solutions containing palladium chloride, ammonium hydroxide, and sodium hypophosphite at 30-70°C.[80]

A new method of precipitating metals such as palladium from aqueous solutions by bombarding metal-salt solutions with high energy electrons (0.5-2 MeV) can be used to produce mirrors, coatings, catalysts, radiographs, and anhydrous compounds.[81]

c. Other deposition methods: Palladium and palladium-alloy films can be readily applied to glass, ceramic, and other nonmetallic surfaces by heating organopalladium compounds, by vacuum deposition or sputtering, or by ceramic-bonded mixtures.

Ceramics and glassware are decorated with palladium by the "liquid-bright method." The "liquid-bright" metal (a varnish-like liquid containing palladium [or platinum or gold] in the form of a complex organic compound) is applied on the substrate and then heated in air at approximately 500°C to decompose the resin and reduce the salt to the metal, which is bonded in a thin layer to the object. Metals can also be used as the substrate.[82]

In vacuum deposition, platinum-group metals, because of their high melting points and low vapor pressures, are vaporized by electron-beam heating or by heating with a helical tungsten wire from a water-cooled copper crucible. The process gives films ≤ 1 μ.[64,83]

Sputtered coatings are produced by dislodging palladium atoms from the metal surface by ion bombardment.[83]

Vacuum deposition, sputtering, or the "liquid-bright" process are used to produce film resistors having very thin films. Thicker film resistors may have screen-printed films.

The films of glaze resistors may contain large metal particles. An example of the latter type of resistor is prepared by applying palladium-silver powders and vitreous flux to porcelain and firing at high temperatures.[73] The use of powdered palladium and palladium-alloy pastes for producing resistors in ceramic hybrid circuits and microcircuits was expected to increase ten-fold within a few years from 1968. The recovery is low in this use.[18,34]

d. Cladding: Palladium is clad on base metals, generally nickel or its alloys, by joining a palladium layer by heat and pressure bonding or brazing to the base metal. The palladium cladding can be precisely located in required areas, which allows the most economical use of the precious metal.[76] When high thermal or electrical conductivity is required, the substrate metal may be copper or silver.[84] A cladding technique is used to produce strip feed stock for punching small parts such as spring electrical contacts and connectors.[76]

V. USES OF PALLADIUM

Palladium, its compounds, and its alloys are widely used for a variety of industrial applications. The average amount of palladium sold annually to industry from 1962 to 1971 was 677,608 troy ounces. The average annual amount used by each industry in that period was: electrical 57.38%, chemical 25.73%, dental and medical 7.55%, jewelry and decorative 2.83%, petroleum 2.52%, glass 0.42%, and miscellaneous and undistributed 3.42%. These industrial uses can generally be divided into noncatalytic and catalytic categories. Specific uses of palladium, listed according to this characterization, are given in Tables A-XIII and A-XIV, pages 74 and 85.

Noncatalytic uses comprise about 70% of all palladium consumed by industry. Sales to the chemical industry, although predominantly for catalytic purposes, probably include palladium used for corrosion-resistant laboratory and process equipment and gas purification units. Brazing-alloy sales are probably dispersed among the electrical, jewelry and decorative, and miscellaneous use figures of the Bureau of Mines.

The next largest use of palladium, after its electrical applications, is as catalysts in the chemical and petroleum industries. Palladium, its alloys, and its compounds are used as catalysts in the production of industrial and research organic and inorganic chemicals, pharmaceuticals, and petroleum products and for a variety of other purposes.

Probably the biggest chemical-industry use of palladium as a catalyst is in hydrogenation, e.g., of intermediates in the production of pharmaceuticals, fragrances, pesticides, dyes, etc., and in the anthraquinone process for producing hydrogen peroxide.[15,69] The use of palladium for catalytic hydrogenation in the chemical industry excluding pharmaceutical applications is described in Section V-B-1. Section V-B-2 discusses hydrogenation in the pharmaceutical industry, and petroleum refining catalysts are discussed in Section V-B-3.

Palladium compounds are also used widely for oxidation and carbonylation in the production of vinyl acetate, acrylic acid, and acetaldehyde (the Wacker process).

Catalytic uses of palladium for hydrogenation, oxidation, carbonylation, oxidative coupling, polymerization, isomerization, fragmentation, disproportionation, and hydrogen transfer reactions are shown in Table A-XIV.

The catalytic mufflers for control of automobile emissions and air pollution control devices other than the catalytic muffler are discussed in Sections V-B-4 and V-B-5, respectively. These devices depend on the oxidation-catalyst activity of palladium and platinum. The catalytic mufflers are expected to use about 100,000 troy ounces of palladium per year until a cheaper, effective alternative system is devised.

A. Noncatalytic Uses

The largest single use of palladium is for contacts in telephone equipment, which prior to 1968, consumed about 300,000-350,000 troy ounces per year, or about 50% of all palladium sold annually in the U.S. Another 8,800 ounces per year (1.3% of total sales) was used as contact material other than in the telephone industry, e.g., in other communication equipment and as electrical components in automobiles, aircraft, and the home.[34,73]

More palladium relay contacts are produced than those of all other metals combined. No other single metal or alloy equals palladium's erosion resistance under mild arcing nor exhibits its relative freedom from troublesome tarnish films.[85]

Palladium electrical contacts, which are limited to use in devices that operate at $\leq$ 5 amp and usually at $\leq$ 1 amp, may be unalloyed palladium, electrodeposited palladium, or palladium alloyed with other platinum metals, gold, copper, silver, nickel, or zinc.[23,82,85]

The uses of palladium and its alloys in electrical contacts, resistance windings and resistors, thermocouples, fuses, fuel cells, brazing alloys, laboratory and process equipment, gas purification, jewelry and decorative applications, the glass industry, dentistry, and medicine are shown in Table A-XIII.

The major gas-purification use of palladium is hydrogen diffusion. Annual palladium consumption prior to 1968 for hydrogen diffusion applications was approximately 7,500 troy ounces ($\sim$ 1.1% of total average annual palladium sales).[34]

Use of palladium in dental alloys probably accounts for almost all of the 7.55% average annual sales attributed to the medical and dental "industry." Dental alloys containing palladium include ultra-high strength alloys, wrought alloys, white alloys, strong age-hardening alloys (alloys that become harder after slow cooling than after quenching), casting alloys and alloys to be covered with porcelain.[73]

The 90% Pt-5% Rh-5% Pd catalyst sometimes used in nitric acid production instead of the preferred 90% Pt-10% Rh gauze is described in Table A-XIV. About 25% of U.S. nitric acid capacity (6.7 million tons produced in 1971) is equipped with new systems to recover lost platinous oxide vapor by forming ternary Au-Pd-Pt alloy with a 20% Au-Pd alloy that is mounted downstream of the Rh-Pt gauze.[87]

B. Catalytic Uses

1. Hydrogenation in the chemical industry and research: Palladium is the best catalyst for hydrogenating olefins and acetylenes and for most hydrogenolysis reactions. It is used for low-pressure hydrogenation of many functional groups such as nitro, nitroso, oximino, cyano, and chlorocarbonyl groups and aromatic systems. When aromatic systems are hydrogenated in the presence of palladium, the aromatic nucleus is not reduced. Palladium can also be used with sodium borohydride and hydrazine. Palladium is useful for high-pressure hydrogenation of phenols and phenol ethers without hydrogenolysis of the hydroxy or alkoxy group.[16,88]

Palladium's virtues as a hydrogenation catalyst include its high activity under ambient conditions, which allows the use of simple, low cost equipment; high selectivity in certain reactions, which gives a product free of by-products; and low reactivity with many reagents so that the product is not contaminated with palladium and palladium can be recovered in high yield from the spent catalyst even under vigorous conditions.[73]

Palladium is generally resistant to poisoning, being slightly deactivated or inhibited by iodide ion, base, and sulfur.[88]

Palladous oxide, palladous chloride, or chloropalladic acid are reduced *in situ* to palladium by hydrogen or other reducing agents to prepare supported catalysts.[17,88] The palladium is dispersed on, for example, pumice; charcoal; or crystalline precipitates of barium sulfate, calcium carbonate, or magnesia.[17] Inhibitors are added to palladium for selective hydrogenation of an acetylene to an olefin; the classic catalyst (Lindlar's) is palladium on calcium carbonate deactivated by addition of palladium acetate and quinoline.[88]

Unsupported palladous oxide, prepared by fusing chloropalladic acid with sodium nitrate, and palladium black, prepared by hydrogenating the oxide or palladous chloride, are less active than supported palladium.[88] Although unsupported catalysts are used less efficiently and mechanical recovery losses are higher, unsupported oxides and blacks are used when adsorption by the support causes substantial product yield losses or poisons the catalyst.[47]

Palladium has also been studied for sterospecific hydrogenations in the form of complexes of poly-S-(amino acids).[89,90] Asymmetric catalysts such as palladium on silk fibroin or d- or ℓ-quartz do not give a product with much optical activity.[88]

Palladium is much more widely used than platinum in industrial heterocatalytic systems, especially in liquid-phase hydrogenation reactions.[78] For selective gas phase reactions, as much as ≥ 40% palladium may be used for special-purpose catalysts. Powdered palladium catalysts, however, usually contain 2-10% palladium on powdered carriers.[73]

Palladium(II) has been used since about 1962 as a homogeneous hydrogenation catalyst. An example of such a reaction is the hydrogenation of dicyclopentadiene in dimethylformamide in the presence of palladous chloride.

2. Hydrogenation in the pharmaceutical industry: Palladium catalysts are used in the hydrogenation of intermediates in the production of such pharmaceuticals as tetracycline, methamphetamine, cyclobarbital, nordefrin, dihydromorphinone hydrochloride (Dilaudid), and meperidine hydrochloride (Demerol).[91] The production of vitamins A and E uses significant amounts of palladium.[15]

A large volume of fermentation produced antibiotics may require several thousand ounces of palladium. The use of palladium as a catalyst for the pharmaceutical industry can be expected to increase with the introduction of new antibiotics. Relatively high losses of palladium occur from these catalysts during handling.[34]

Lederle uses a 5% palladium-on-carbon catalyst in the production of its tetracycline antibiotic Achromycin.[92] Another commercial method of producing tetracycline is by direct fermentation by *Streptomyces aureofaciens*.[93] Trace amounts of palladium are always found in Achromycin when palladium catalyst is used to dechlorinate chlortetracycline. A sample of nonmedicinal tetracycline hydrochloride was found to contain 2.8-ppm palladium.[94,95] Since the usual human daily dose of tetracycline is 2,000 mg, palladium consumption could be 0.0056 mg/day.

Numerous other pharmaceuticals may be prepared by catalytic hydrogenation of intermediates. These include certain amphetamines; dihydromorphine; dihydrostreptomycin; dihydrotachysterol; ephedrine; epinephrine, isoproterenol, and levarterenol from catechol; estriol; folinic acid-SF; menthol from thymol; methyprylon; oxytocin; phenindamine tartrate; and thyroxin.[95,96] The specific catalysts used are not readily identified, but probably contain palladium in many cases.

3. Petroleum refining: Palladium catalysts are used in petroleum refining chiefly for hydrocracking (hydrogenation plus pyrolysis) of distillate fractions[39] with some use for reforming of cracked petroleum fractions into high-octane gasoline,[9] hydrodesulfurization (hydrotreating),[97] and hydrogenation refining. Annual purchases vary because of initial fills and 2-4 year use cycles.[34]

A big fixed-bed hydrocracking process (Unicracking-JHC, developed by Union Oil and Jersey Standard) uses molecular-sieve catalysts supplied by Union Carbide, two of which are presumably based on palladium and two are based on nonnoble metals such as nickel and tungsten.[87] Catalyst cost is estimated at 2-7¢ per barrel of fresh feed. In late 1972, the total hydrocracking capacity utilizing this process was estimated to be 370,000 barrels per stream day.[98]

Other hydrocracking processes are Isomax (a fixed-bed process by UOP and California Standard), which presumably uses a precious metal on an amorphous support, and processes by Gulf, Shell, and American Oil. American Cyanamid and Harshaw are major suppliers of hydrocracking catalysts.[87]

The molecular-sieve catalysts contain ~ 0.5% noble metal, usually platinum or palladium dispersed as atoms in a natural or synthetic zeolite. The catalyst life of zeolites before regeneration can be as long as 100 days or more in the absence of metal contaminants in the feed.[99]

Other hydrocracking catalysts comprise the noble metal deposited on a support such as silica-alumina.[99]

4. Catalytic mufflers: Automotive exhaust emissions control devices represent the largest potential area of short term growth for palladium. The use of these noble-metal containing devices, in an attempt to meet the 1975-1976 automobile emission standards, may require as much as one-third of the total platinum-group metals demand in 1976.[37]

The rapid development of catalytic mufflers is the result of the Clean Air Amendments Act of 1970. The original law required a 90% reduction in carbon monoxide and hydrocarbon automobile exhaust emissions from 1970-model levels by the year 1975, and a 90% reduction in nitrogen oxides from the 1971-model levels by the year 1976. Interim levels set by the Environmental Protection Agency for carbon monoxide, hydrocarbons, and nitrogen oxides for 1975 are 15.0, 1.5, and 3.1 g/mile, respectively. The State of California has established more stringent standards for 1975: 9.0, 0.9, and 2.0 g/mile, respectively. The 1976 United States standards were set at 3.4, 0.41, and 2.0 g/mile, respectively, and, in addition, the acceptable nitrogen-oxides level was to be reduced to 0.4 g/mile in 1977.[100,101] However, in December 1973, the Senate approved the extension of 1975 auto emission standards through 1976.[102]

Identification of the best methods of control and the likelihood these requirements will have to be met are uncertain at this time. Techniques used for measuring nitrogen oxides levels recently have been found to be in error. New information now indicates that only the atmospheres of Chicago and Los Angeles have hazardous levels of nitrogen oxides.[103] The Environmental Protection Agency has subsequently requested authority to establish nitrogen oxides limits administratively.

Controversy has arisen concerning the oxidation of the sulfur in gasoline (0.01 to 0.10%) into sulfur dioxide, in the presence of automobile exhaust catalysts to produce sulfuric acid mist and sulfates.[103-105] Atmospheric sulfate concentrations are reportedly generated that are significantly higher than those at which asthma and heart disease sufferers are bothered. The catalysts cause an increase of about 0.05-g sulfates per mile.[106] The heart of the controversy centers around whether or not such emissions would effect a significant increase in the sulfur dioxide levels in urban atmospheres. Approximately 99% of the sulfur dioxide currently in the air is produced by sources other than automobiles, such as power plants.

In addition to these uncertainties, there are no comprehensive data available on the metallic emissions from these control devices. The Environmental Protection Agency has launched a $1.5 million research program to answer questions on metallic particles and other nonregulated exhaust emissions.[107]

In March 1973, the Committee on Motor Vehicle Emissions of the National Academy of Sciences reported that delay in implementation of the 1970 amendments until 1977 would allow development of more promising technology and not lessen air quality appreciably.[108] If such a recommendation were included in a special study on auto guidelines due to be issued next summer by the Academy, the immediate future of these devices would be further complicated.

The 1970 law also requires that emission systems be guaranteed for 50,000 miles of driving. None of the Big Three automakers has been able to meet the 1975 standards for 50,000 miles consistently.[109] Ford's system for 1975 eliminates 92% of the hydrocarbon and 81% of the carbon monoxide in the exhaust. Model 1973 cars have been able to reduce the emissions by about 80% and 70%, respectively.

A control system capable of reduction as well as oxidation is required to meet the emission standards for carbon monoxide, hydrocarbons, and nitrogen oxides. This type of system will be required to meet 1977 standards and apparently will utilize different catalysts from those to be used through 1975.

Research and development effort directed toward development of catalytic systems capable of hydrocarbon oxidation has been ongoing for many years. Work prior to 1946 showed that platinum; palladium; oxides of copper, cobalt, and nickel; iron chromate; and catalysts promoted by vanadium or manganese are suitable for oxidizing hydrocarbons and carbon monoxide.[110] Co_3O_4, Cr_2O_3, and $CuCr_2O_4$ were found to be the most active transition-metal oxides. Platinum and palladium activities per unit metal weight, however, may be 100 to 1,000 times greater than transition metal compounds under exhaust conditions.[111]

General Motors has tested some 100 different catalyst formulations submitted by 60 different companies and has generally found the durability of transition base metal catalysts to be poor.[112]

Small amounts of palladium enhance base metal catalyst properties. Effective pelleted transition metal catalysts containing < 0.05 ounce palladium have recently been developed with lifetimes of 50,000 miles in simulated laboratory aging tests. Promising base metal catalysts have been developed so recently that vehicles equipped with them have not yet accumulated high mileage.[111]

Noble metals achieve higher conversion efficiencies at lower temperatures than base metal catalysts, have higher thermal stability on oxide pellets, a common support; and are less susceptible to sulfur poisoning.[111]

Supported platinum-metal catalysts have high activity for reducing nitrogen oxides. Their selectivity for nitrogen as the reduction product rather than ammonia, which can be reoxidized, is improved by adding other noble or base metals. Platinum or palladium, for example, can be used with nickel or other good ammonia-decomposition catalysts.[111] Ruthenium, osmium, and rhodium have also been studied as reducing catalysts. Of these, ruthenium appears to be the most promising because of its efficiency and lower tendency to produce ammonia.[113]

Converters capable of controlling nitrogen-oxide emissions, however, are much less developed than oxidation converters. Since a single catalytic converter for nitrogen oxides, hydrocarbons, and carbon monoxide is not possible until significant carburetion improvements insuring the proper air-fuel ratio are made, two-stage (dual) converters will probably be used to meet 1977 standards. These dual systems will reduce nitrogen oxides in the first catalyst section and oxidize the residual hydrocarbon and carbon monoxide in the second.[111]

General Motors Corporation, Ford Motor Company, and Chrysler Corporation will all utilize various types of platinum-group metal catalysts in their 1975 models. A summary of the available information on the devices to be used is given in Table A-XV, page 92.

General Motors will use three different converters containing porous alumina pellets approximately 28.194 mm in diameter coated thinly with 0.050 oz (1.4 g) platinum and palladium. General Motors catalysts reportedly have an approximate platinum-to-palladium ratio of 7:3 or 5:2. The larger converters will contain about 2.7 kg (6 lb) of pellets.114/

The underfloor converters, which will be on 98% (90-100%114/) of General Motors 1975 cars, comprise four stampings of a new stainless steel containing 11% chromium and no nickel. The stampings include two "louvers," which retain the catalyst, and two identical shells, which form a pod around the catalyst retainer. The four stampings are stacked and then joined by electron-beam welding and by two or four heavy stainless steel studs. The converters are insulated with ceramic-fiber wool and wrapped in aluminum.115/

The General Motors converters operate at 482-806°C (900-1500°F) with temperatures of ≤ 982°C (1800°F) tolerable for 1 or 2 min.115/

A dual-mode emission control system had been predicted for some 1975 General Motors models, in which an exhaust diverter or by-pass protected the catalyst and canister from damage during high temperature operations.113/ However, as of September 15, 1973, by-passes apparently were not to be included.

General Motors has contracted to buy 120,000 troy ounces of palladium per year from 1974 to 1983 as well as 300,000 troy ounces of platinum per year from Impala Platinum, Ltd., Johannesburg, South Africa.37/

W. R. Grace is to supply General Motors with 12-15 million pounds of catalysts per year in 1975, 1976, and 1977. Grace will produce these catalysts at its Davison Chemical Division's plant in Curtis Bay, Maryland, beginning in mid-197437/ and will utilize some of American Cyanamid's technology.101/ The other three General Motors suppliers--Engelhard, of Newark, New Jersey; Air Products, of Calvert City, Kentucky; and an American Cyanamid-Japan Catalytic International partership--will each supply 8-10 million pounds of catalyst per year. The Cyanamid catalyst (Aeroban) sold to General Motors will contain platinum and palladium and will be produced at Azusa, California.116/

General Motors is to fabricate its converters (6.5 million per year) and stainless steel housings at its AC Spark Plug Division plant near Milwaukee and, perhaps, at Flint, Michigan.[137]

American Motors is to buy most of its converter assemblies from General Motors.[114]

For controlling emissions of nitrogen oxides from 1976, General Motors has tested dual catalytic converters in which the nitrogen oxide-reducing catalyst was located in the exhaust pipe at the exhaust manifold outlet. The catalyst for oxidizing hydrocarbons and carbon monoxide was mounted under the floor pan. Dual-bed arrangements containing both oxidizing and reducing catalysts in the same canister have also been studied.[113]

Ford is expected to equip about 65% of its 1975 and 1976 models with converters.[114] Ford is obtaining its catalyst systems from Engelhard (60%) and Matthey-Bishop (30%).[117] Matthey-Bishop will produce catalysts in its Devon, Pennsylvania, plant.[109] The catalyst systems are promoted platinum metals on monolithic honeycomb ceramic supports, which have parallel channels for gases and are shaped to fit inside a stainless steel reactor the size and shape of a conventional muffler. The converter systems are located under the toeboard.

All V-8 Ford engines will have two converters except those of Maverick, Comet, Econoline, and Bronco models. All other Ford, General Motors, and Chrysler models have one converter per car (see Table A-XV).

At the end of 1973, Chrysler decided to put converters on almost all of its 1975 cars.[102] Chrysler is to obtain 100,000 oz of palladium from the USSR. Universal Oil Products was to produce catalysts for Chrysler in Tulsa, Oklahoma, beginning deliveries by March 1974,[111,117] but UOP decided to purchase catalyst material for the 1975-1977 model years from Engelhard.[118] The converters, which are monolithic and located underfloor, reportedly contain < 0.1 troy ounce platinum and palladium in 5:2 ratio.[109] Universal Oil Products will also supply pellet catalysts to Japan's Nissan, Toyota, and Daihatsu Motor Companies.[101]

Engelhard and Matthey-Bishop will apply the ceramic slurry and install the catalyst on the substrate, usually a ceramic compound comprising magnesium oxide, aluminum, and silicon, produced by Rhone Progil, American Lava, Corning, and W. R. Grace and Company.[101,109] Muffler producers, such as Walker Manufacturing and Arvin Industries, Inc., will place the monoliths in containers of stainless steel or other alloys. The ceramic pieces are protected by wire mesh in most converters.

Most catalyst deterioration in Ford's tests occurred in the first 10,000 miles. Ford's and Chrysler's entire catalytic canisters will have to be replaced, but General Motors will replace only the pellets.[109]

Noble-metal catalysts have a smaller volume, which is also a consideration for Chrysler and Ford; their cars lack the room for bigger converters. Ford and Chrysler chose monolithic over pellet converters chiefly because of their smaller under-car space.101/

Engelhard will also supply converters to Toyota and Nissan Motor companies of Japan for their 1975-1978 models to be exported to the U.S. Engelhard Kali-Chemi A.G. will supply the Engelhard device to Sweden's Volvo for its 1975-1977 models.119/ The total capacity of Engelhard's plants in Newark, New Jersey, and Huntsville, Alabama, is about 4 million units per year.101/ The platinum-palladium ratio of the catalysts is reportedly 2:1.

There are a number of significant problems yet to be overcome in the day-to-day operation of these catalytic devices. Catalysts can be destroyed by additives in the fuel such as lead or phosphorus; temperatures above 980-1095°C (1800-2000°F) will melt the ceramic-monolith substrate [in the 1975 systems, the catalytic reaction temperature range is 174-1045°C (350-2000°F)]; and excessive vibration will pulverize the brittle ceramic substrate.109/

Although catalytic converters require unleaded (≤ 0.03 g/gal) gasoline, General Motors tests indicate that the converters may not be damaged by an occasional tankful of leaded fuel.112,120/

If large amounts of unburned fuel pass through the engine and are "ignited" in the converters, a near explosion of the catalyst often occurs. Running out of fuel, a fouled carburetor, two or more defective spark plugs, dieseling, backfiring, and possibly turning off the ignition while the car is underway, can cause this. The 1975 models will be equipped with various features to avoid these occurrences.109/

United States regulations stating that catalytic systems must be effective for 50,000 miles are interpreted as allowing for one catalyst replacement at 25,000 miles.100/ General Motors has reported to the Environmental Protection Agency that its platinum-palladium catalytic converter will last the lifetime of a car or "more than 50,000 miles." However, these catalysts cannot meet the 1976 interim standards.112/ All of the several hundred catalysts tested by General Motors were reported to have deteriorated and disintegrated in use.121/ However, General Motors President Edward N. Cole told the Senate Public Works Committee in November 1973, that loss of noble-metal particulates was only a problem during the early stages of converter engineering.122/

Shelef and Gandhi of Ford Motor Company found that use of platinum (Universal Oil Products catalyst) and palladium (Engelhard) is hampered by carbon monoxide formation and ammonia formation.[123]

These catalysts are theoretically capable of emitting metallic oxide or carbonyl pollutants. Metal containing condensation nuclei were induced at 185-800°C in filtered air, in a 3% carbon monoxide-nitrogen mixture, and in the products from a pulsed-flame combustor from catalysts containing chromium, copper, and nickel. These conditions are similar to those of catalytic automotive emissions control systems. Start-up and idle operations, which give 2-8% carbon monoxide concentrations and 300°C exhaust temperatures, as well as acceleration following idle periods and high-speed cruising, giving 1000°C temperatures, would be conducive to such metallic emissions.[124]

When a catalyst is destroyed, the car's performance will not be affected, but emissions will rise immediately. An over temperature warning light is provided by some manufacturers to warn drivers of the fact.[109]

Auto engineers feel that some sort of mandatory inspection (possibly every 5,000 miles) will be necessary to insure that owners will maintain their cars much better than they have in the past.[109]

5. Catalytic air pollution control other than catalytic mufflers: In 1968, catalytic air pollution control systems used 6,000 ounces of palladium annually. Palladium catalysts are preferred in the oxidation of organic industrial emissions to reduce nitrogen oxides and oxidize sulfur compounds. To minimize pressure drop, the commercial catalysts for air pollution control are generally precious metals applied to stainless steel screens, porcelain rods, or honeycomb ceramic.[125] For example, El Paso Products Company eliminated yellow fumes from the stack gas of their nitric acid plant by installing a tail gas treatment system containing a palladium-coated ceramic honeycomb catalyst, which catalyzed the low temperature reaction of the nitrogen oxides with any commonly available fuel.[92,126] Operating temperatures of these systems range from 350 F for ethylene combustion to 750° and 800°F for asphalt and methane fumes, respectively.[125] Palladium has no potential substitute in this use.[12]

6. Miscellaneous catalyst uses: Catalysts containing palladium are patented for hydrogen peroxide decomposition,[127] heavy-water production,[1 and oxygen scavenging in containers for oxygen-sensitive products such as foods.[129-131] Oxygen can be removed from hydrogen or other gases containing sufficient hydrogen by converting to water in the presence of palladium on alumina.[76]

Currently, Engelhard is trying to sell a process that involves palladium for purifying contaminated (e.g., river) water.[33]

VI. LOSSES TO THE ENVIRONMENT

Losses of palladium to the environment result from two general categories of activity: that associated with the production and fabrication of palladium and palladium-containing products and that resulting from the use and ultimate disposal of these materials. The following discussion summarizes the data available on palladium losses to the environment as the results of these activities.

A. Production and Fabrication Losses

Palladium is a minor constituent of the platinum-group metals recovered in the Goodnews Bay District of Alaska. Between 1936 and 1967, 84.5% of the metals was platinum; 11.4%, iridium; and only 0.39%, palladium.2/ The chemical form of palladium in this deposit is not well defined. Most of the platinum metals in this alluvial deposit occur as fine metallic grains. The pay streak consists largely of clay. This clay is often so cohesive that it does not disintegrate readily in the trammel screen of the dredge or in the succeeding gigs and riffles. Hence, a substantial loss in platinum metals occurs during dredging. Studies indicate that reworking of waste from these dredging operations is profitable.

The chemical properties and the physical form of the platinum metals in the Goodnews Bay deposit are such that they should pose no environmental or human health problems.

Platinum metals are recovered as a by-product of gold placer mines in California.2/ These placers contain predominately platinum and osmiridium generally as an intergrown alloy. The palladium content of this ore, based on crude assay results, appears to average less than 1%. Hence California placer mines, which have been estimated to have produced only 30,000 troy ounces of platinum metals between 1887 and 1963, have resulted in little environmental contamination by palladium. Again this chemical form of palladium is not easily solubilized and should pose no environmental problems.

The other major source of placer-mined platinum metals comes from Oregon.2/ It is estimated that 1,500 ounces of platinum metals have been recovered. Analysis of two samples of this ore showed palladium accounts for about 0.2% of the platinum metals present.

Little information can be found on even the total platinum metal content of other ores such as copper and nickel from which precious metals are recovered as a by-product. However, the smelting conditions are commonly designed for good platinum metal recovery. It is estimated that about 1 oz of precious metal is recovered for each 35 tons of refined copper produced or for each 6,000 tons of copper sulfide ore. The total palladium produced in the United States in 1970 from domestic sources--crude platinum and gold and copper refining--was 11,851 oz.[18] It is estimated that 90% of the precious metal in the Sudbury ore is recovered. This is probably a higher percent recovery than is achieved from U.S. copper ores, since 16% of copper is fire-refined and palladium is only recovered from copper that is electrolytically refined.

Again, however, losses from the mining of copper would appear to have no detrimental environmental consequences.

Beneficiation and smelting of platinum-metal concentrates is described in Section IV-B. Possible points of loss during these operations are difficult to identify.

The losses of palladium in the original ore occurring during beneficiation of copper ores cannot be estimated; however, such losses would involve palladium in its natural chemical form and should pose no environmental hazard.

During smelting of copper and other base-metal ores containing platinum metals, the amount of palladium lost is dependent on what elements must be removed from the base metal. In copper refining, the palladium loss could be as high as 10-50% if iron must be eliminated.[12] The predominant loss would involve retention of palladium by the slag, which should pose no problem. Other possible routes of loss include vaporization and entrainment of palladium particulates in gases contacting the ore during the various stages of smelting.

The temperatures the ore concentrate encounters during smelting vary from 1200 to 2400°F. The vapor pressure of palladium at 2400°F is about 0.001-mm Hg, which is about five orders of magnitude higher than that of platinum, the next most volatile platinum-group metal.[132] Consequently, some palladium, the most volatile of the platinum-group metals, could be lost during these operations. Palladium loss during copper smelting would be extremely difficult to determine since the metal is alloyed with the copper and the volatility of palladium in such alloys has not been determined. However, one can make a crude estimate of the loss of palladium by volatilization during smelting using the following assumptions:

1. The palladium-copper solution behaves as an ideal solution.

2. The mole fraction of palladium in copper is 4.5×10^{-7} (0.5 ounce per 35 tons of copper).

3. The rate of vaporization of pure palladium at 1315°C (2400°F) is 0.3×10^{-9} g/cm^2-sec.[132/]

A calculation using Henry's law indicates the rate of palladium vaporization would be 1.35×10^{-16} g/m^2-hr.

If one made the additional assumptions that the copper that contains palladium is subjected to a temperature of 2400°F for 2 hr and that the ore being smelted (1,720,000 tons in 1970) is 0.1 m deep, one calculates that the total surface available for vaporization is 5,000,000 m^2 for 2 hr. Based on the above assumptions, the loss of palladium during smelting should be less than 0.1 g/year. Although this is a very crude estimate, it is probably within two to three orders of magnitude of the correct value and indicates little loss of palladium occurs by vaporization during copper smelting.

Data is available on palladium content of the particulates collected from International Nickel Company smelter stacks in Sudbury (Table A-XVI). These data show that palladium can be lost in gaseous and particulate emissions from smelting operations. The losses would appear to arise from palladium-particulate entrainment in the flue gas streams rather than from palladium vaporization. Using the numbers reported for the palladium content of particulates from the INCO copper stack and using the 1969 reported particulate losses from U.S. copper smelters of 8,700 tons,[133a/] one calculates a possible annual palladium loss to the atmosphere from U.S. copper smelters of about 1,200 troy ounces. This number may be too high since the Sudbury ore has a higher initial platinum-metal content. The error should be offset in the opposite direction, however, by the greater effort of INCO to recover the platinum metals.

No more than 0.25% of platinum metals contained in Sudbury concentrates (obtained by electrolyzing a secondary anode and working up the slimes) is lost by the specialized Mond Platinum Metals Refinery in England.[23/] All precious metal bearing dust, fumes, residues, and solutions in such refineries are collected, recycled, and retreated.[133b/] However, losses in the U.S. precious metal recovery processes from used and new sources is probably closer to 5%. Much used and new palladium goes through the copper cycle (losses from Sudbury ores to refined platinum metals are ~ 10%[23/].)[a/] This would tend to offset the higher yields from recovering palladium from petroleum refining and other catalysts and from electroplating solutions (98-100%). Engelhard Industries claims that 95% of the platinum-group metals can be recovered from scrap.[30/]

---

a/ Losses of new palladium in the U.S. from copper ores may be higher since copper recovery is only 83%.[8/]

It can be calculated that domestic precious-metal smelters and refiners may have lost about 4,000 troy ounces of palladium annually in the period 1967-1971 with the assumptions that 16% of the toll-refined palladium was from crude foreign platinum (and the loss rate was only 0.25% or 1,500 troy ounces from that source) and that the remainder of toll-refined, new, and used palladium was lost at a 5% rate to give a total loss of about 220,000 troy ounces in the 5-year period. This 44,000 troy ounces loss would be distributed among minor losses to the atmosphere by volatilization, losses to water from discarded liquid wastes, losses incorporated in slags and other solid wastes, and losses incorporated in other refinery products in trace amounts.

If the losses from primary and secondary refineries were entirely attributed to losses to water at 44,000 troy ounces per year or about 120 troy ounces per day, and if the total effluent stream from all refineries is 1 million gallons per day, these streams could carry as much as 1-ppm palladium (7 ppm is toxic to the fish Orizias latipes).[134]

Because of its expense, care is exercised in the handling of palladium during fabrication, and most that is lost mechanically as shavings, etc., is returned in the sweeps to secondary refiners.[12]

Palladium is probably not lost by vaporization or oxidation during fabrication or alloy preparation.[12,33] Any apparent loss of palladium in melting operations is due to impurities in the original palladium weighed.[11]

Following is a summary of palladium losses during mining, ore beneficiation, smelting and refining, and fabrication and their apparent environmental consequences.

1. Losses of palladium to the environment during mining of alluvial deposits is likely to be a substantial percent of the metal present in the original ore (≥ 10-17%). The palladium in the mining debris would be in the same chemical form as in nature. No information was found that ties any local health problems to natural palladium contamination in areas where alluvial deposits occur. Palladium in alluvial deposits occurs as an alloy of platinum metals. Because of the resistance of these alloys to corrosion, as evidenced by their survival for millions of years in these chemical forms, they should pose little or no environmental hazard.

Palladium in copper ores is believed to occur in conjunction with the mineral sperrylite (platinum arsenide) or as a very dilute solid solution of the major ore minerals, probably as substituents in the crystal lattices.[2] Although reliable solubility data is not available, it does not appear that residual palladium in copper mine tailings would leach out even when residual sulfur in the tailings is oxidized to create an acidic condition in a tailings pile.

2. Losses of palladium during smelting of copper and other base metal ores can be substantial depending on the ore. Losses of 10-50% apparently occur when the copper ore contains excessive amounts of iron.12/ Much of this palladium is retained in the slag. The palladium retained in such slag should be in very low concentrations (probably < 1 ppb) and should pose no health hazard.

Other modes of loss during smelting would include vaporization and palladium-particulate entrainment in smelter flue gases.

Calculations of losses by vaporization, although very crude, indicate that losses by this route are very small even if the values are in error by two to three orders of magnitude.

Losses by particulate entrainment in base metal smelter flue gases appear to be substantially higher than losses by vaporization. At present there are no data available for estimating losses by this route. (The palladium content of particulate emissions from the INCO copper stack in Sudbury, Canada, is about 8 ppm.) Any loss would probably be as submicron particles.

3. Loss of palladium during primary and secondary smelting and refining may be as high as 44,000 troy ounces per year or about 1,500 troy ounces per refinery per year. It is possible that wastewater streams leaving such refineries could contain potentially hazardous quantities of palladium.

4. Losses during fabrication are probably negligible because of the extreme care exercised in collecting wastes.

B. Losses from Use and Disposal

Limited data are available on which to estimate losses of palladium resulting from product use and disposal. Some generalizations can be made, however, about the losses from a few major use categories.

1. Catalyst losses by petroleum refining and chemical industry: In petroleum refining, loss of palladium in the form of hydrocracking catalysts through fines losses, etc., may be similar to values estimated for platinum reforming catalysts: 0.3%111/ or ~ 2%.12/ It can be calculated that an average annual amount of 50-340 troy ounces of palladium may have been lost from this source in the 1962-1971 period.

It would appear from the wide variations in purchases by the petroleum industry that large purchases reflect mostly newly installed hydrocracking catalysts and that much of the catalysts are effectively recovered in toll refining.

On the other hand, the smaller fluctuation in purchases by the chemical industry in the period 1966-1971 (the largest purchase, in 1968, is only 24% larger than the smallest purchase in 1967), may indicate that most of the palladium being purchased by the chemical industry represents make-up for for catalyst losses and a small percent (perhaps 10%) for growth (expansion of processes on stream) or new uses. The chemical industry probably uses far greater amounts of palladium each year than it purchases, as reflected by the amounts annually toll refined, at least half of which are probably owned by the chemical industry. (Little can be inferred from totaling the annual amounts toll refined, since much palladium probably has been recycled more than once within any given period.)

Reportedly, platinum used in the chemical and petroleum industries is recycled at a rate of ≥ 85%.[135] It probably has not been particularly economical to recycle palladium catalysts at such a high rate. In 1963, the producers' prices for platinum and palladium were $77-$80/troy ounce and $26-$28/troy ounce, respectively; in 1969, these prices were $130-$135/troy ounce and $43-$45/troy ounce.[8] Prices in August 1973, were $158-$163/troy ounce platinum and $84-$86/troy ounce palladium.[136] In January 1974 these prices were $170-$175/troy ounce and $90-$92/troy ounce, respectively.[137] Possibly, the much higher relative price of palladium will stimulate further recycling of this metal.

Because of the diversity of catalytic uses for palladium and its compounds, it is extremely difficult to determine from the open literature how much catalytic palladium is used. Industrial sources will not identify how much palladium is used for particular processes, and sellers of palladium and its compounds will not identify the purchasers and amounts sold to them. We can only make a crude estimate that roughly 90% of the average annual purchase by the chemical industry is make-up for lost catalyst: 160,000 troy ounces.

How much of this is lost as metallic or alloyed palladium, which is probably innocuous unless contaminating the air, and how much is lost as palladium(II) solutions from homogeneous-catalyst waste streams could not be determined. The latter use is relatively large and is involved in the production of acetaldehyde (625,000 tons annual capacity at Bay City, Bayport, and Longview, Texas) and vinyl acetate (318,000 tons annual capacity at Bay City, Bayport, and Houston, Texas).[138,139] The production of acrylic acid and analogous compounds by carbonylation of α-olefins probably also uses significant amounts of palladium chloride.[140,141] The broad geographic distribution of other chemical plants that use palladium catalysts makes difficult the identification of specific potential sources of return of the metal to the geosphere and the pathways of such return (air, water, or soil).

Small amounts of palladium are lost in catalyzing the oxidation of ammonia to nitric acid. Each daily ton of nitric acid capacity uses 2 troy ounces of catalyst gauze (a minor amount of which is a 90% Pt-5% Rh-5% Pd alloy; the preferred catalyst is a 90% Pt-10% Rh alloy). The gross consumption of the gauze is 0.008 troy ounce per ton of nitric acid (53,600 troy ounces per year). The 20% Au-Pd alloy mounted downstream of the catalyst gauze can reduce the platinum loss to 0.004 oz/ton.87/ Since only 25% of U.S. capacity is so equipped, this would reduce the annual loss of platinum metals in this use to 46,900 troy ounces, and the total loss of palladium from use of the minor catalyst would be far less than 2,000 troy ounces. It is claimed that 75% of the platinum-group metals used in nitric acid production is recovered.37/

2. Losses to the environment from catalytic mufflers: The current catalytic mufflers are not designed for recyclability. The palladium and platinum in them will not be recoverable after the junked auto is baled and sent to the scrap metal converter.12/ However, a National Materials Advisory Board Report issued in March 1973111/ assumed that 75-90% of the platinum (and, presumably, palladium) will be collected for recyling. This rate appears to be optimistic in the light of current catalyst attrition rates. J. W. Dunham, General Manager of the Automotive Products Division of Univeral Oil Products, predicts that U.S. demand for platinum and platinum and palladium would be cut considerably if large-scale salvage is developed. He claims that 90% of the salvaged noble metals can be reclaimed by processes for recovering platinum from petroleum-refining catalysts.118/

Potential losses of palladium to the environment during operation of the catalytic muffler are ill defined at this time. Preliminary data, however, indicate that 30-40% of the catalyst in candidate emission control devices will be lost during a life cycle of 50,000 miles.142/ These emissions are presumed to be in the form of fine, particulate metal, oxides, or sulfides.

It is presently estimated that 10 million vehicles will be equipped annually with such devices and that a muffler's average life will be approximately 5 years (50,000 miles).143/ Using an estimate of 40% catalyst attrition during the useful life of the catalyst and up to half of this attrition occurring during the first year of operation and assuming that each catalytic muffler contains 0.01 troy ounce of palladium, one calculates that 8,000-20,000 troy ounces of palladium would be lost from this source in 1975 and that the annual loss will approach 40,000 troy ounces by 1979.

The remainder of the palladium in spent mufflers will likely be discarded to the environment until such time as platinum and palladium prices or the number of spent mufflers justify recovery. Estimated value of residual platinum and palladium in spent mufflers will approach $33 million annually by 1979 based on current metal prices.137/

Whether this magnitude of palladium loss occurs will depend on whether adequate substitutes for the catalytic muffler are found and whether the current proposed automobile exhaust emission limits remain in effect, which is in some doubt at the time of this writing.

Ultimately, automakers will probably turn to less effective base metals that may last only 10,000 miles compared with the longer lives of palladium and platinum and yet still be cheaper.[12] General Motors is committed to the catalytic converter system for at least 1975 and 1976; an alternative system reportedly cannot be introduced in less than 3-4 years.[112] Presumably, other U.S. automakers are in the same position.

3. Attrition of palladium during contact use: The life of palladium contacts is about 10 times the life of silver contacts with respect to erosion and sticking. However, small amounts of organic vapor in the operating atmosphere of a palladium contact decrease its lifetime if the contact opens and closes currents. Sources of such vapors in telephone contacts may be organic filler material in the relay. Organic vapors may also arise from coil forms, wire coatings, insulation, soldering flux, and potting and sealing compounds. When the contact arcs slightly, most organic vapors will deposit carbon; this in turn, activates the contact, thereby increasing the arcing time and increasing the erosion rate of the contact metal, particularly that of the cathode. Activated palladium contacts always give cathode arcs, confining the metal loss largely to the cathode. Metal is removed from the anode by short arcs, and during contact separation, by metal that may have melted to form metal bridges. In the absence of contamination, little metal is lost, but undesirable shapes may form, leading to sticking. In the presence of organic vapors, a dust formed from carbon mixed with metal vapors is lost.

Excessive organic vapor must be avoided in "dry contacts," which open and close without arcing and do not carry current or make or break, since solid polymerized organic material may build up to a thickness sufficient to cause occasional transient failures to close a circuit.[69,73,144] Approximately 75% of all contacts in Bell Telephone switching systems do not operate in arcing conditions and hence do not erode.[144] These benefit from a 1-mil Au-Ag overlay to prevent polymer formation.

The main loss of palladium in electrical uses, however, would not be contact attrition, but rather the lack of recovery from junked appliances. If it is believed that almost all of palladium used in telephone equipment is recovered and the remainder of the palladium in electrical uses is too dissipated for recovery, the amount lost is calculated to be 39,000 troy ounces.

4. Dental: Some palladium alloys used in dentistry can be recycled: orthodontic appliances, and, probably, worn dentures and bridges when they are replaced. The total average annual amount of palladium used in dentistry and medicine is 51,236 troy ounces. How much of this amount is "lost" (that is, will probably never by recycled) in gold alloy cast inlays, crowns, and other tooth restorations, was not determined. Possibly, as more people can afford the best dental care, this use may increase significantly (see Table A-XIII), and even more palladium will never by recycled. On the other hand, dental alloys are being developed with significantly reduced precious metal contents.145/ In any event, the palladium in question is already in direct human contact and assumed to be innocuous in the form of dental alloys. Therefore, it will not be considered in estimating the total amount of palladium "lost to the environment."

5. Losses from other uses: Palladium brazing alloys are used to join base metals, ceramics to metals, and precious metals. Only in the latter use would any palladium ever likely be recycled. The unreclaimed metal in junked apparatus would be so widely dispersed as not to pose any hazard.

No problems of disposal would be expected from the use of palladium jewelry. Processing scrap and old jewelry are carefully recycled.

The glass industry uses metallic palladium in small amounts that will probably never be recycled (see Table A-XIII, page 82), but, again, disposal and loss constitute little hazard because of the wide dispersal of the instruments containing very small amounts of palladium.

C. Summary of Losses

We estimate that the average annual loss of palladium to the domestic environment in recent years has been about 244,000 troy ounces distributed among losses to the atmosphere during copper smelting, 1,200; losses during primary and secondary refining, 44,000; petroluem refining, 50-340; electrical uses, 39,000; and chemical-industry catalysts, 160,000 troy ounces. In addition, each model year of cars equipped with catalytic mufflers containing palladium will add about 8,000-20,000 troy ounces of palladium per year to the environment, which will amount to about 40,000 troy ounces in 1979 when approximately half of the U.S-manufactured automobiles are expected to be so equipped.

VII. PHYSIOLOGICAL EFFECTS OF PALLADIUM AND ITS COMPOUNDS

Palladium and its compounds apparently do not pose serious environmental or industrial health problems of the type associated with many platinum-group metals. Threshold limit values for palladium or its compounds have not been established. Skin sensitization, such as that caused by platinum, is not generally atributed to palladium.[96,146]

Human intake of palladium compounds either by subcutaneous injection, topical application, or oral ingestion has not caused serious acute systemic effects. No adverse effects have been reported in humans from long-term exposure to palladium or its compounds.

However, palladium chloride was fatal when injected into rabbits and to the fish Orizias latipes when present in the water. Chronic ingestion of palladium chloride in concentrations of 7 ppm has caused cancer in mice and rats.

Palladium(II) ions are even more toxic to microflora, being of toxicity comparable with the very poisonous ions of, for example, silver, copper and mercury. In a higher plant (Poa pratensis), 10 mg/liter palladium chloride was toxic.

On the molecular level, palladium(II) forms complexes with amino acids and proteins, inhibits or inactivates enzymes, and degrades DNA.

The physiological effects of palladium compounds are discussed more fully in the following subsections.

A. Human Effects

Palladium metal, palladium chloride, and palladium hydroxide have been evaluated for their therapeutic effects. Colloidal palladium has been used to treat tuberculosis and gout. Such treatments were found to be ineffective, and were innocuous or caused a feverish reaction.[147] Colloidal palladium has also been used to treat obesity. Injection of 5-mg palladium into the abdominal wall was found to cause fever, weight loss, and a black-and-blue coloration at the injection site that persited for 2.5 years but gave no albuminuria or glycosuria.[146]

Palladium chloride was long (and ineffectively) used to treat tuberculosis at the dosage of about 18 mg/day.[147] Oral dosages of 65 mg/day apparently produced no adverse effects. Topical applications of palladium chloride as a germicide is not known to have caused any skin irritation.

Palladium hydroxide has also been used to treat obesity. Colloidal palladium hydroxide injections (5-7 mg/day) reportedly caused 19-kg weight loss in a 3-month period with necrosis at the injection site.[147] Dilutions of 1:25,000 showed a hemolytic effect.

Palladium is not a factor in platinosis syndrome.[146] The only report found of contact dermatitis from palladium referred to a research chemist who had been studying various precious metals for several months. His face, hands, and arms showed patches of eczema. This condition cleared up completely on avoidance of palladium exposure and treatment with betamethazone valerate ointment.[148]

Palladium has been found in teeth containing palladium-alloy fillings. Presumably, small amounts of palladium are solubilized by body fluids. Traces of palladium have also been found in human liver.[149]

B. Animal Effects

Palladium chloride (0.2 cm^3/200 g body weight i.v.) has a slight diuretic activity in rats.[150] Orally ingested palladium is excreted chiefly by the feces, but intravenous injections of palladium hydroxide or palladium chloride are fairly rapidly excreted in the urine.[146]

Feeding 5-ppm palladium (as $PdCl_2$) in drinking water from birth to death suppressed the growth rate of male and female mice. Palladium fed males survived longer than controls. Palladium was slightly carcinogenic to the mice[151] and was carcinogenic to rats fed small doses of palladium for life.[152] Administering 5-mg palladium per day to rabbits for 2 months, however, did not cause any health disturbances.[153]

An Ag-Pd-Au dental alloy imbedded subcutaneously for 504 days caused tumor formation in 7 of 14 rats.[154] Implants of an Ag-Pd-Au-Cu dental alloy (70.02, 24.70, 5.23, and 0.03%, respectively), in the oral submucous membranes of rabbits gave only mild effects, but liver implants in rats temporarily constricted capillaries of the liver parenchyma, and testicular implants in rats caused seminiferous-tubule degeneration.[155]

The lethality of intravenous injections of palladium chloride in rabbits has been shown to be dose-rate dependent; 1.7 mg/kg was fatal to a rabbit on the 17th day after intravenous injection. Palladium caused hemolysis and albuminuria, acted as a diuretic, and was eliminated chiefly through the kidneys. The heart, kidneys, liver, bone marrow, and blood cells were the chief sites of damage. Palladium was also detected in the lungs, spleen, and muscle.[147]

Four hours after intravenous injection of ^{103}Pd as Na_2PdCl_4 at a dose of 0.5-2 μCi/rat, 60% of the ^{103}Pd had been eliminated in the urine. Significant amounts of ^{103}Pd were retained by the kidney, liver, and spleen at 7 days; and the urine contained 76% of the administered ^{103}Pd and the feces, 13%: liver and kidney retained detectable amounts of ^{103}Pd 16 days after injection. The half times for removal of ^{103}Pd from the kidney and liver were 9 days and 6 days, respectively.[149]

When injected subcutaneously, palladium chloride is unabsorbed; all but one of 24 rats given 4-24 mg/kg buffered palladium chloride survived unimpaired.[147]

Intramuscular injection of ^{102}Pd (~ 50 mg/2 ml solution) gave a mild lymphocytic reaction in rabbits. An inert palladium black suspension in gelatin injected into the omentum of rabbits gave a similar, slight inflammatory reaction. Both palladium solution and suspension remained well localized at the injection site.[153]

The complexes ^{109}Pd-hematoporphyrin and ^{109}Pd-protoporphyrin were recommended for use in selective lymphatic ablation to control homograft rejection. The complexes concentrated in the liver and lymph nodes of dogs rather than other abdominal organs, lung, muscle, or bone marrow. However, rabbit spleens accumulated the complexes in higher concentrations. The hematoporphyrin complex caused bleeding abnormalities.[157]

6-Mercaptopurine (mp) and its palladium complex $Na_2[Pd(mp)_2Cl_2]$ had the same leukopenia-induction capacity when given orally to 5-day-old chicks at equal dosage levels for 10 days; the palladium complex was less toxic.[158] $Na_2[Pd(mp)_2Cl_2]H_2O$ and $[Pd(butp)_3Cl]Cl$ where butp = butylthiopurine were extremely active against Adenocarcinoma 755 and Sarcoma 180 in mice. $[Pd(da)_2Cl_2]$ and $[Pd(dcda)Cl_2]_2$ (where da = daraprim [2,4-diamino-5-(4-chlorophenyl)-6-ethylpyrimidine] and dcda = dichlorodaraprim) were inactive as anticancer agents.[159]

The 24-hr lethal concentration of palladium chloride to a cyprinodont fish (*Orizias latipes*) was 0.00004 *M* (7 ppm).[134]

C. Plant Effects

Approximately 1 ppm was the median effective toxic dose of Pd^{2+} for the fungus Alternaria tenuis (osmium, mercury, and silver ions were more toxic).[160]

The growth of brewers' yeast was reduced 50% by 2.7-ppm palladium; 10-ppm palladium inhibited growth completely. Palladium was ranked among the very poisonous ions of cadmium, copper, silver, osmium, and mercury.[161]

Palladium chloride in amounts 10^{-5}, 10^{-4}, and 10^{-3} molar added to the medium fluid (deionized water) of Schizophyllum commune in a study of the effects of trace elements on L-malate fermentation through a carbon dioxide-fixing process decreased the amount of glucose consumed to 91%, 51%, and 30%, respectively, of that consumed in the basal medium and decreased the amount of L-malate produced to 75%, 25%, and 4%, respectively. Palladium chloride enhanced the accumulation of pyruvate, possibly due to inhibition of the carbon dioxide-fixing process. Of the ions tested, including Pb(II), Cr(III), Co(II), Ni(II), Cd(II), Sn(II), V(V), Se(IV), and Pd(II), the last two were the most inhibitory.[162]

Growth of Kentucky bluegrass (Poa pratensis) was stimulated by palladium chloride until toxic concentrations were reached. The phytotoxicity may have been due to a nonspecific ionic effect. Palladium was detected in the roots but not the leaves.[163] In nutrient solutions, $>$ 3 mg/liter palladium chloride inhibited transpiration of Kentucky bluegrass; 10 mg/liter concentrations were toxic. Palladium chloride caused aberrant stomatal histogenesis; inhibited modal meristem development; changed chloroplast structure; and caused hypertrophy of mesophyll cells, nuclei, and nucleoli.[164]

D. Molecular Biology

Palladium(II) forms complexes with L-cysteine, L-cystine, and L-methionine in solution but not with L-histidine. Pd^{2+} binds the proteins carboxypeptidase, casein, papain, and silk fibroin.

α-Chymotrypsin and trypsin are inactivated by Pd^{2+}, Cu^{2+}, Zn^{2+} and Hg^{2+}; catalase, lysozyme, peroxidase, and ribonuclease were not inactivated by Pd^{2+}. Possibly, Pd^{2+} inactivates trypsin by combining with free -SH and (or) cystine groups and inactivates chymotrypsin by reacting with cystine groups.[165] Chloropalladate ions are toxic to tyrosinase.[149] At $\leq$ 2 mg/ml, palladium chloride significantly inhibited lactic dehydrogenase of the white sucker fish Castostomus commersoni (the toxicity of Pd^{2+} was higher than that of Hg^{2+} or Ag^{+}). Inhibition of glutamic oxalacetic transaminase was also high but less than that by Hg^{2+}, Ag^{2+}, or CN^{-}.[166]

Pd^{2+} incubated with calf-thymus DNA degraded the DNA molecule. In a medium of low ionic strength, Pd^{2+} had low degradation activity.[167]

VIII. HUMAN HEALTH HAZARDS

Major point sources of environmental contamination by possibly hazardous forms of palladium are primary and secondary smelter and refiners of copper, lead, silver, gold, and other precious metals. Many of these have been identified in Table A-XII, page 72. Minor amounts of palladium and palladium oxide would be lost from these sources as vapors and submicron size particulates. The concentrations are apparently so low (e.g., Table A-XVI, page 93, INCO Emissions) that human health hazard is unlikely from palladium alone.

Automobiles equipped with palladium containing catalytic mufflers disseminate aerosol palladium or palladium oxide, a possible health threat to those inhaling metallic particles in urban atmospheres. No information was found regarding the health effects of inhaling particulate palladium metal. By 1979, when ~ 50% of all cars are expected to be equipped with catalytic mufflers, these catalytic devices may spew as much as 40,000 troy ounces palladium into the atmosphere. This value is comparable with that calculated for palladium lost by all pathways to the environment from smelting and refining operations. Certainly, the potential for harm is greatest in urban population centers where refiners are concentrated, such as in New Jersey.

Losses of palladium used as industrial catalysts could not be verified since relative amounts used as supported and unsupported palladium and as palladium(II) compounds were not determinable. Losses are probably significantly lower for supported palladium catalysts, and small amounts of metallic palladium lost in handling probably pose no health threat in themselves. We have estimated that 160,000 troy ounces of palladium may be lost annually by the chemical industry, a significant amount of it in the form of divalent species.

Refinery liquid wastes and untreated catalytic and electroplating solutions discharged to streams by users too small to consider palladium recovery economical may, in some cases, contain amounts of palladium(II) toxic to aquatic life forms, particularly fish and microflora. Reports of accumulation and concentration of palladium by organisms were not found. Apparently, the possibility of biological magnification of palladium in man's food chain is remote.

Direct oral ingestion of palladium(II) in contaminated water and plants containing concentrations sublethal to them may be hazardous. In the past, doses of 65-mg palladium chloride per day have caused no immediately discernible adverse effects in human tuberculosis patients.[147] However, the fact that chronic oral ingestion of palladium chloride in rats and mice increased the incidence of cancer raises the question of its possible carcinogenicity in humans.[151,152]

IX. CONCLUSIONS AND RECOMMENDATIONS

A. Conclusions

1. Palladium and its compounds do not cause any recognized acute or chronic toxic affects in humans. However, the carcinogencity of palladium(II) compounds in rats and mice as well as the toxicity of these compounds to mammals, microflora, a fish (*Orizias latipes*), and a higher plant (Kentucky bluegrass) are cause for environmental concern.

2. We estimate the average annual loss of palladium to the domestic environment in recent years has been about 244,000 troy ounces, a significant amount of which comprised the innocuous palladium metal or palladium alloys.

3. A significant part of the estimated 160,000 troy ounces per year of palladium(II) lost by the chemical industry must be in the form of palladium(II) compounds in wastewater streams. We were unable, however, to quantify the amount of palladium(II).

4. Loss of palladium or palladium oxide particulates to the atmosphere from the operation of automobiles equipped with platinum-palladium catalytic mufflers may be as much as 40,000 troy ounces in 1979, when approximately half of U.S.-manufactured automobiles are expected to be so equipped. This value is comparable with the approximately 44,000 troy ounces per year estimated as the losses of palladium to the environment by all pathways during primary and secondary refining.

5. Loss of palladium(II) compounds by refiners and electroplaters to wastewater streams may be sufficiently concentrated to be toxic to aquatic organisms.

6. The loss of palladium or palladium oxide vapors or particulates to the atmosphere during copper smelting is estimated to be about 1,200 troy ounces per year. The concentration of palladium in the emissions is thought to be too low to pose any hazard even in the vicinity of a smelter.

7. The major use of palladium and its alloys is in the electrical industry. For the purposes of this report, it was estimated that the major portions of these purchases for use in telephone equipment is ultimately recovered and that the 39,000 troy ounces used for other purposes is too widely disseminated to be recovered or pose a health hazard.

8. Losses of palladium metal or alloys during use as petroleum-refining catalysts is assumed to be minor (50-340 troy ounces per year).

9. Palladium metal or alloy losses to the environment were assumed to be innocuous and minimal from use for brazing alloys, dental alloys, jewelry, and glass coatings and from fabrication of palladium and its alloys (except in the case of electroplating wastes).

B. Recommendations

Epidemiological studies in areas where palladium losses to the environment are high (e.g., near precious-metals refineries) might determine if palladium is a genuine carcinogen in humans or causes other previously unrecognized health problems.

Any health hazards arising from the drastically increased concentrations of palladium in urban atmospheres from the catalytic muffler will probably also be obscured by symptoms produced by platinum, whose soluble salts cause the syndrome of respiratory and skin disorder "platinosis."

Although, apparently, environmental contamination by palladium poses no urgent problem to humans, its toxicity to lower life forms makes regular palladium-loss appraisals and increased recycling of palladium wastes worthy undertakings.

If any link to human diseases should be found for any particular form of palladium, the problem of determining the relative amount of this form that is lost by the chemical industry should be resolved. Perhaps extensive mailings to users of palladium catalysts (if a statistically meaningful number could be identified) in the light of new knowledge regarding palladium toxicity would yield the desired information.

APPENDIX A

TABLES CITED IN THE TEXT

LIST OF TABLES

Number	Title
A-I	Concentration of Palladium in Various Sources
A-II	Natural Alloys and Intermetallic Compounds of Palladium
A-III	Natural Palladium Compounds with Oxygen, Sulfur, Tellurium, Arsenic, Antimony, Bismuth, and Tin
A-IV	Imports of Refined Palladium
A-V	Imports for Consumption of Palladium, Unwrought; Partly Worked; or as Bars, Plates and Sheets ≥ 0.125 In. Thick
A-VI	U.S. Refinery Production of Palladium
A-VII	General Services Administration Stock of Palladium
A-VIII	U.S. Palladium Stocks at Year End, Troy Ounces
A-IX	U.S. Exports of All Platinum-Group Metals except Platinum
A-X	Palladium Sales by Industry
A-XI	Palladium Supply and Demand, 1968
A-XII	Palladium Metal Works
A-XIII	Noncatalytic Uses of Palladium
A-XIV	Catalytic Uses of Palladium
A-XV	Comparison of 1975 Catalytic Converter Systems
A-XVI	INCO Smelter Stack Emissions, 1971

TABLE A-I

CONCENTRATION OF PALLADIUM IN VARIOUS SOURCES

Source	Palladium Content (ppm)	Reference
Crystal abundance	0.01-0.02, 0.013	1, 168
Ultramafic rocks (dunites, etc.)	0.12	169
Mafic rocks (basalts, gabbros)	0.019	170
Basalts (basic rock)	0.02, 0.041	1, 169
Diabases (basic rock)	0.016	1
Granites (acid rock)	0.001-0.01	169
USA silicic rocks	0.003	1
Bushveld silicic rock	0.005	1
Bushveld gabbros	0.031	1
Other gabbros	0.020	1
Bushveld norites	0.015	1
Sudbury norites	0.006	1
Bushveld peridotites	0.012	1
Other peridotites	0.011	1
Bushveld pyroxenites	0.059	1
USA pyroxenites	0.003	1
Igneous rocks	0.01, 0.17, 0.02	1, 169
Average meteorite content	1.4	1
Carbonaceous chondrites (meteors)	0.500-0.810	3
Russian Au-containing deposits	0.05-3	171
Arctic watershed soils and sediments	0.007-0.0035	172
Chromitite zones Stillwater Complex, Montana, Average	0.96	173
East-Central Alaska nickeliferous serpentinite	< 0.004-0.011	174
Canadian uranium ores	Trace to 443	175
Canadian nickeliferous sulfide ores	0.03-246.8	175
Canadian nonnickeliferous sulfide ores	Trace to 1.03	175
Sudbury chalcopyrite, footwall	0.96	176
Sudbury pyrite	1.08	1
Sudbury pyrrhotite	1.04	1
Sudbury pentlandite	4.07	1
Sudbury chalcopyrite	6.21	1
Sudbury maucherite	24.8	2
Noril'sk, USSR, pyrrhotite	9.98-15.0	1
Noril'sk chalcopyrite	3.74-4.60	1
Canadian Ni matte	3.4	177
Canadian Cu refiners Doré metal	1,020	178

TABLE A-II

NATURAL ALLOYS AND INTERMETALLIC COMPOUNDS OF PALLADIUM[1]/

Mineral	Description	Wt % Pd
Allopalladium	Hexagonal. Pd with some Pt, Ru, and Cu and traces of Ir, Rh, Ag, and Hg.	
Avaite	Synonym of platiniridium.	
Eugenesite	Synonym of allopalladium.	
Gold	Reported to contain Pd.	Maximum reported 11.6
Iridosmine-osmium series	Hexagonal.	0-0.6
Norilskite	A mixture (?) of Pt, Fe, Ni, and Cu.	3.6
Palladium	Cubic, Pd with some Pt, Rh Ir, Ru, Os, and Pb.	86-100
Palladium amalgam	Synonym of potarite.	
Palladiplatinum	Cubic. Pd and Pt.	51.1
Platiniridium	Cubic. Ir and Pt.	9.9
Platinum	Cubic.	Usually 0-2 Maximum 3.
Porpezite	Palladian gold.	$\leq$ 10
Potarite	Pd_3H_2 or PdHg(?). Cubic (?).	
Selenpalladium	Synonym of allopalladium. Erroneously thought to contain Se.	
Unnamed mineral A	Pd_4Pb	$\approx$ 70

TABLE A-III

NATURAL PALLADIUM COMPOUNDS WITH OXYGEN, SULFUR, TELLURIUM, ARSENIC, ANTIMONY, BISMUTH, AND TIN[1/]

Mineral	Composition	System	Remarks
Arsenopalladinite	Pd_3As	--	--
Braggite	(Pt, Pd, Ni) S	Tetragonal	--
Cooperite	(Pt, Ni, Pd) S	Tetragonal	--
Froodite	$PdBi_2$	Monoclinic	--
Hollingsworthite	(Rh, Pt, Rd)-(As, $S)_2$	--	
Kotulskite	Pd (Te, $Bi)_{1-2}$	Hexagonal	
Michenerite	$PdBi_2$	Cubic	From Sudbury, Ontario
Moncheite	(Pt, Pd)(Te, $Bi)_2$	Hexagonal	--
Palladinite	PdO(?)	--	Uncertain
Rhodian sperrylite	(Pt, Rh, Ir, Pd)-(As, $S)_2$	--	--
Stannopalladinite	(Pd, $Pt)_3Sn_2$	--	--
Stibiopalladinite	Pd_3Sb	--	--
Vysotskite	(Pd, Ni, Pt)S	Tetragonal	--
Zvyagintsovite	(Pd, $Pt)_3PbSn$	--	--
Unnamed mineral G	Pd(Sb, Bi)	--	--
Unnamed mineral H	Pd-Bi mineral	--	From Sudbury
Unnamed mineral I	Pd_2CuSb	--	--
Unnamed mineral J	Pd_3CuSb_3	--	--
Unnamed mineral L	PtPdSn	--	--

TABLE A-IV

IMPORTS OF REFINED PALLADIUM[179/]

Year	Amount (troy oz)
1963	503,843
1964	483,018
1965	734,881
1966	985,137
1967	737,082
1968	1,168,511
1969	632,172
1970	773,817
1971	657,983

TABLE A-V

IMPORTS FOR CONSUMPTION OF PALLADIUM, UNWROUGHT; PARTLY WORKED; OR AS BARS, PLATES, AND SHEETS ≥ 0.125 IN. THICK

Country	1968[180]	1969[180]	1970[29]	1971[180]	1972[180]
Canada	78,550	27,994		18,479	12,449
Mexico	494				
Finland		200			
Sweden	350			888	
Denmark		325		400	
Norway	4,410	5,638	2,100	5,840	14,565
United Kingdom	417,620	253,570	106,615	254,643	190,425
Ireland	6,936				
Netherlands	63,017	25,410		631	
West Germany	4,828	1,849	2,129	1,612	6,840
Belgium	1,830	731			
France	15,292				
Switzerland	26,196	8,425	27	458	2,098
USSR	419,543	269,875	389,031	332,909	487,618
Spain	10,669	16,076			
Australia					974
Japan	17,860	8,112	502		23,558
Republic of South Africa			1,952	39,163	93,470
Other			1,929		
Total Imports[29, 180]	1,087,595	618,205	503,783	657,983	831,997
Total Imports[179]	1,168,511	632,172	773,817	657,983	

TABLE A-VI

U.S. REFINERY PRODUCTION OF PALLADIUM, TROY OZ [9, 15, 18, 29, 59, 179, 181, 182]

	New				Total New	
Year	Domestic Sources	Foreign Sources	Total New	Used	and Used	Toll Refined
1961			28,988	32,451	61,439	
1962	14,141	2,003	16,144	56,273	72,417	
1963	28,099	4,700	32,799	59,993	92,792	
1964	22,863	4,438	27,301	49,879	77,180	326,000
1965	22,500	3,839	26,339	50,025	76,364	203,000
1966	29,907	1,460	31,367	50,009	81,376	~ 657,000
1967	8,142	120	8,262	215,162	223,424	708,966
1968	5,275	83	5,358	195,620	200,978	1,055,470
1969	8,224	163	8,387	227,763	236,150	945,106
1970	11,851	24	11,875	208,555	220,430	569,711
1971			20,951	161,099	182,050	593,842 a/
1972				175,000		

a/ 84% used, 16% from foreign

TABLE A-VII

GENERAL SERVICES ADMINISTRATION STOCK OF PALLADIUM 7, 8, 18, 179, 182, 183/

Year	End of Year, troy oz	Change from Previous Year
1961	745,819	+100,009
1962	745,819	0
1963	737,935	-7,884
1964	737,935	0
1965	737,935	0
1966	837,491	+99,556
1967	908,760	+71,269
1968	1,002,760	+94,000
1969	1,081,949	+79,189
1970	1,249,832	+167,883
1971	1,329,021	+79,189

TABLE A-VIII

U.S. PALLADIUM STOCKS AT YEAR END, TROY OZ.

Year	Refiners', Importers' and Dealers' Stocks[29, 179, 181]	New York Mercantile Exchange[29, 59]
1961	244,910	
1962	285,173	
1963	315,756	
1964	317,691	
1965	427,450	
1966	574,651	
1967	460,624	
1968	393,882	306,500
1969	608,716	117,500
1970	332,726	32,700
1971	316,126	21,600

TABLE A-IX

U.S. EXPORTS OF ALL PLATINUM-GROUP METALS EXCEPT PLATINUM[7, 59]

Year	Amount (troy oz.)
1961	20,460
1962	10,940
1963	11,776
1964	21,167
1965	30,172
1966	103,425
1967	118,267
1968	172,159
1969	277,495

TABLE A-X

PALLADIUM SALES BY INDUSTRY, TROY OZ. 7, 18, 29, 34, 179, 181/

Year	Chemical	Electrical	Dental and Medical	Jewelry and Decorative	Petroleum	Glass	Miscellaneous and Undistributed	Total
1961	90,533	353,010	47,228	14,354	449	5	2,471	508,050
1962	110,518	327,788	54,899	12,975	961	124	12,585	519,850
1963	118,757	331,868	42,940	13,880	16,008	20	3,054	526,527
1964	117,102	350,889	49,893	20,886	41,887	110	10,665	591,432
1965	156,796	430,384	50,192	18,203	37,001	1,402	23,107	717,085
1966	221,559	531,545	67,102	32,215	28,760	1,011	12,020	894,212
1967	192,011	324,684	56,085	18,676	3,506	301	25,878	621,141
1968	228,318	329,012	61,636	17,797	22,683	10	62,023	721,479
1969	214,508	430,258	52,326	21,837	1,337	3,891	34,581	758,738
1970a/	184,618	429,032	47,583	17,329	15,494	21,147	24,140	739,343
1971a/	218,651	431,505	61,594	18,752	2,916	237	26,451	760,106
1972								876,024
1973	124,200				1,400			
2000b/	460,000	450,000	100,000	30,000	80,000		105,000	1,225,000

a/ Revised 1971.
b/ Contingency forecast 8/

TABLE A-XI

PALLADIUM SUPPLY AND DEMAND 1968, TROY OZ.[8/]

Imports	1,156,000
U.S. Refined (New)	11,000
U.S. Secondary Refined	196,000
Toll Refined	1,055,000
Industry Stocks, Jan. 1, 1968	461,000
Total Supply	2,879,000
Industry Stocks, Dec. 31, 1968	394,000
Toll Refined	1,055,000
Exports	155,000
Other Stocks	462,000
U.S. Stockpile Acquisition	94,000
U.S. Industrial Sales	721,000
Total Demand	2,881,000

TABLE A-XII

PALLADIUM METAL WORKS, U.S.A.[179, 184-186]

Company	Plant Site	Activity
Inspiration Consolidated Copper (Anaconda associate)	Inspiration, Miami, Ariz.	Electrolytic Cu refiners
Nagma Copper Co.	San Manuel, Ariz.	Electrolytic Cu refiners
Sel-Rex Co., Div. of Oxy Metal Finishing Corp.	Chatsworth, Calif.	Precious-metal plating and secondary refining
Electronic Space Products, Inc.	Los Angeles, Calif.	Refiners
Martin Metals, Inc.	Los Angeles, Calif.	Refines and manufactures metals, chemicals and salts. Secondary refiners.
Western Alloys Refining Co.	Los Angeles, Calif.	Secondary refiners[a]
Handy and Harman	Los Angeles, Calif.	Secondary refiners
American Smelting and Refining Co. (ASARCO)	Selby, Calif.	Pb refiners
PGP Industries, Ind., subsidiary of Gerald Metals, Inc.	Santa Fe Springs, Calif.	Refiner[181]
Wildberg Bros. Smelting and Refining Co.	South San Francisco, Calif.	Refiners
The Wilkinson Co.	Westlake Village, Calif.	
The J. M. Ney Co.	Bloomfield, Conn.	Fabricators
Handy and Harman	Fairfield, Conn.	Secondary refiners
American Chemical and Refining Co., Inc.	Waterbury, Conn.	Refiners
E. I. du Pont de Nemours & Co., Inc.	Wilmington, Del.	Producers of Liquid Bright for ceramics and electronics
Kahlenberg Labs, Inc.	Sarasota, Florida	Producers of colloidal Pd suspended in aqueous medium
Goldsmith Division of National Lead Co.	Chicago, Ill.	Fabricators, chemicals
M & G Metal Products, Co.	Chicago, Ill.	Fabricators
Simmons Refining Co.	Chicago, Ill.	Secondary refiners
D. F. Goldsmith Chemical and Metal Corp.	Evanston, Ill.	Producers of Pd and $PdCl_2$
United Refining and Smelting Co.	Franklin Park, Ill.	Refiners
Mercer Refining	Melrose Park, Ill.	Refiners
Deringer Metallurgical Corp.	Mundelein, Ill.	Fabricator
Joseph Behr & Sons, Inc.	Rockford, Ill.	Secondary refiners
American Smelting and Refining Co. (ASARCO)	Baltimore, Md.	Electrolytic Cu refiners
D. E. Makepeace, Div. of Engelhard Minerals and Chemicals Corp.	Attleboro, Mass.	Fabricators
Texas Instruments, Inc.	Attleboro, Mass.	Fabricators
Eastern Smelting and Refining Corp.	Lynn, Mass.	Primary and secondary refiners
Kennecott Copper Corp.	Ann Arundel County, Md.	Electrolytic Cu refiners
Cerro Copper and Brass, Div. of Cerro Corp.	St. Louis, Mo.	Electrolytic Cu refiners (scrap)
The Anaconda Co.	Great Falls, Mont.	Electrolytic Cu refiners
American Smelting and Refining Co.	Omaha, Neb.	Pb refiners
J. Wm. Krohn Co., Inc.	Carlstadt, N.J.	Refiners and fabricators
United States Metals Refining Co., subsidiary of American Metal Climax, Inc.	Carteret, N.J.	Electrolytic refining of Cu ores and scrap. Secondary refiners[31]
Am. Platinum and Silver Div., Engelhard Industries, Inc.	Newark, N.J.	Fabricators and refiners
Baker Platinum Div., Engelhard Industries, Inc.	Newark, N.J.	Produces metal, alloys, fabricated products
Engelhard Industries, Div., of Engelhard Minerals and Chemicals Corp.	Newark, N.J.	Primary and secondary refiners
Hudsar, Inc. (formerly Hudson Smelting and Refining Co.)	Newark, N.J.	Smelts residues containing Pt
Sel-Rex Co., Div. of Oxy Metal Finishing Corp.[b]	Nutley, N.J.	Precious metal plating and secondary refining.
International Smelting and Refining Co. (Anaconda)	Raritan, Perth Amboy, N.J.	Electrolytic Cu refiners[c]
Electronic Development Corp.	Roseland, N.J.	Fabricators of Pt [Pd?]
Spiral Metal Co., Inc.	South Amboy, N.J.	Fabricators of Pt products and dental alloys [Pd?]

TABLE A-XII (Concluded)

Company	Plant Site	Activity
Metz Refining Co.	South Plainfield, N.J.	Refiners
H. A. Wilson Div., Engelhard Industries	Union, N.J.	
Matthey Bishop, Inc.	Winslow, N.J.	Refiners
HPM Div., Aremco Products, Inc.	Briarcliff Manor, N.Y.	Producers of single crystals
Ciner Chemical Refining Co., Inc.	Brooklyn, N.Y.[d/]	Refiners
Hoover and Strong, Inc.	Buffalo, N.Y.	?
Williams Gold Refining Co., Inc.	Buffalo, N.Y.	Au refiners
Atomergic Chemetals Co., Div. of Gallard-Schlesinger Chemical Manufacturing Corp.	Carle Place, L.I., N.Y.	Refiners and fabricators
MacKenzie Chemical Works, Inc.	Central Islip, N.Y.	Producers of Pd acetylacetonate
Sherman Industries, Am. Silver Co. Div.	Flushing, N.Y.	Fabricators
Phelps Dodge Refining Corp.	Laurel Island, L.I., N.Y.	Electrolytic Cu refiners
J. Aderer, Inc.	Long Island City, N.Y.	?
Handy and Harman	Mt. Vernon, N.Y.	Fabricators and secondary refiners?
Johnson, Matthey and Co., Inc.	New York, N.Y.	Refiners and fabricators
Leico Industries, Inc.	New York, N.Y.	Refiners and fabricators
United Mineral and Chemical Corp.	New York, N.Y.	Produces high-purity Pd and its compounds
Nassau Smelting and Refining Co., Inc., Div. of Western Electric	Tottenville, Staten Is., N.Y.	Secondary refinery (telephone scrap)
Secon Wire, Div. Secon Metals Corp.	White Plains, N.Y.	Fabricators
Cincinnati Gold and Silver Refining	Cincinnati, Ohio	Au and Ag refiners
The Shepherd Chemical Co.	Cincinnati, Ohio	Producers of Pd acetylacetonate and Pd octoate
Matthey Bishop, Inc. (J. Bishop & Co. Platinum Works)	Malvern, Pa.	Fabricators, chemical producers
Semi-Elements, Inc., subsidiary of Riker-Maxson Corp.	Saxonburg, Pa.	Producers of single crystals
Geo. E. Conley Co., Inc.	Providence, R.I.	Secondary refiners
Pease and Curran, Inc.	Warwick, R.I.	Refiners
Technic Inc.	Providence, R.I.	Produces Pd for electroplating
American Smelting and Refining Co. (ASARCO)	El Paso, Texas	Cu and Pb refiners
Phelps Dodge Refining Corp.	El Paso, Texas	Electrolytic Cu refiners
American Smelting and Refining Co. (ASARCO)	Tacoma, Wash.	Electrolytic Cu refiners
Kennecott Copper Corp.	Garfield, Utah	Electrolytic Cu refiners (Anode slimes toll-refined by others)

a/ Advertisement, Secondary Raw Materials, 10 (4), 130 (1972).
b/ Advertisement, Secondary Raw Materials, 10 (2), 142 (1972).
c/ Treated slimes from the Great Falls plant are refined at the Raritan Copper Works.[32/]
d/ Advertisement, American Metal Market/Metalworking News, 80 (191), 41 (1973).

TABLE A-XIII

NONCATALYTIC USES OF PALLADIUM

ELECTRICAL USES

Electrical Contacts

Use	Form of Palladium	Remarks	References
Electrical relays such as in telephone equipment for low-noise contact service with extreme reliability.	40% Cu-60% Pd alloy or 40% Ag-60% Pd alloy	At ~400°, these alloys form thin oxide films that decompose at higher temperatures, discouraging sticking in light-duty electrical contacts. Ag-Pd alloys containing >50% Pd have max. resistance to sulfidation or ozone oxidation, but those with ~40% Ag have max. hardness and resistivity. Alloys containing >60% Pd are used for contacts operating at low-contact forces.	17 66 73 82
Relays on cable circuits where considerable shunt capacity exists and in telegraph repeaters.	Western Electric Company W.E. #4 (40% Cu alloy)	The 40% Cu alloy has max. hardness. Usually used in conjunction with a contact or in manufacture of slip rings.	66 73
Bell Telephone relays.	One contact is solid Pd; the other, 22 karat Au-Pd alloy plate.	Most resistant combination to erosive wear, sliding damage, and friction-polymer formation.	187
Ford Motor Co. automobile alternator regulators.	Pt-Pd-Ag allcy.		92
Make-and-break contacts for choppers, tape-readers, and automotive signal flashers; sliding contacts for switching disks, cermet trimmer pots, and potentiometer windings; telephone spring wire relays; brushes for switches and wires in brush assemblies; and ductile raw material for parts with severe forming.	Pt-Pd-Ag-(Au) alloys.		188
Deposit on the Ni-Fe alloy reeds of reed switches.	Pd (or Rh or Ru).	Used to replace gold where contact sticking or welding occurs.	189
Slip rings running against brushes of the same material.	40% Cu-60% Pd		69
Electromagnetic relays in control systems, engineering equipment, lighting, and appliances.	Pd	Less preferred than Ag and Ag alloys.	69

ELECTRICAL USES: Contacts (Continued)

Use	Form of Palladium	Remarks	References
Electromagnetic vibrators for automotive voltage regulators, bells, buzzers, horns, and radio vibrators.	Pd-Ag-Ni-W	Second choice; W-0.5 Mo preferred.	69
Thermomechanical thermostats in household heating and cooling, cooking, and electric blankets.	Ag-Pd	Second choice; Ag preferred.	69
Precision snap-action switches in heat thermostats.	35% Pd-30% Ag-14%. Cu-10% Pt-10% Au-1% Zn.	Fewer than 1% of these heat thermostats use this alloy; Hg-Mo or Hg-Pt preferred.	69
Slip rings and brushes in sliding contacts of telecommunications app.	Pd-Ag-Cu	Pure Ag, 90% Ag-10% Cu, or 70% Au-39% Ag preferred.	69
Sensitive relays in telecommunications.	Pd	Pd preferred to Au-Ag-Pt or Pt.	69
General-purpose relays in telecommunications app.	Pd	Pd usually preferred.	69
Switches in telecommunications app.	Pd		69
Armature-type sensitive relays of micro-contacts.	Pd and Au-plated Pd.		69
Micropotentiometer brushes.	60% Pd-40% Cu and 35% Pd-30% Ag-14% Cu-10% Pt-10% Au-1% Zn.	Au-Ag-Pt and Au-Ni-Zn-Cu preferred.	69
Wire of micro slip rings.	70% Au-16% Cu-7% Ag-5% Pt-1% Pd-1% Zn.	Pt-Rh alloy preferred.	69
Brush rings (micro slip rings).	35% Pd-30% Ag-14% Cu-10% Pt-10% Au-1% Zn.	Au-Ni-Zn-Cu and Au-Ag-Pt alloys preferred.	69
Barrier deposits under gold to prevent gold's diffusion into the underlying metal, e.g., silver or tin-lead solder.	Pd (or Rh).		70 189
Printed circuit switch segments.	Pd (or Rh).		189
End connectors of printed circuit boards.	Electrolytically deposited Pd coatings.		76
Spring elec. contacts and connectors.	Pd clad to a base metal or electrolytic deposits of Pd.		76

TABLE A-XIII (Continued)

ELECTRICAL USES (Continued)

Resistance Windings and Resistors

Use	Form of Palladium	Remarks	References
Resistance windings in potentiometers; electronic components in heating pads, voltage regulators, thermostats, and artificial-horizon instruments.	Pd alloys, e.g. Au-Pd-Fe resistance wire.		1 69 76
Potentiometers.	40% Ag-60% Pd alloy, Pd-Au alloys akin to dental alloys, Mo-Pd alloys.	The Ag alloy has an unusually low temp. coefficient of resistivity.	17 73 190
Capacitors, resistors, other thick-film electronic components.	Pd and Pd-alloy films, e.g. Au-Pd-Pt resistance films.	Applied to glass, ceramic, and other non-metallic surfaces by the "liquid-bright" process, vacuum deposition, or sputtering (thin films, e.g. Pd-Au alloys). Thicker films are applied by screen printing [e.g. of inks containing Pd black (Pd. PdO)] or by fusing powders (e.g. of Ag and Pd) and a vitreous flux to porcelain.	69 73 78 191
Printed circuits.	Pd plating.	Used as an alternative to gold.	77 78
Resistors in ceramic hybrid circuits and microcircuits.	Powdered Pd and Pd-alloy pastes.		18 34

Thermocouples

Use	Form of Palladium	Remarks	References
Thermocouples in laboratory furnaces and thermocouple-sensing elements for monitoring high-temperature areas of turbo-prop jet engines (Allison Division of General Motors).	Pt-Pd-Au alloy.	Metallurgical industries, metal-physics laboratories, and the missile and aerospace industries are the main users of such thermocouples.	7
	(Engelhard's Platinel thermocouple: a 60% Au-40% Pd alloy as the neg. leg and the pos. leg is Pd-Pt-Au.)	Platinel is among the most widely used Pt-metal alloys for temperature measurement. The Platinel thermocouple has lasted for 4 months in an application where a base-metal thermocouple was replaced daily: 2370°F (1299°C) in a both oxidizing and reducing atmosphere.	76 92 192

TABLE A-XIII (Continued)

ELECTRICAL USES (Continued)

Thermocouples (Concluded)

Use	Form of Palladium	Remarks	References
	Platinel II (a 65% Au-35% Pd alloy as the neg. leg and a 14% Au-31% Pt-55% Pd alloy pos. leg).	Preferred to the Pd-60% Au thermocouple. Useful at ≥1300°C. Produces same emf as the Chromel-Alumel couple.	72 73
Thermocouples with high emf other than the Platinel.	Au-Pd/Pt-Ir and Pd/Pt-Ir.		73
Low-temperature (room temperature to 1000°C) thermocouples.	Pallador Pd alloys.		73
Small-diameter thermocouples for use at < 1000-1100°C.	40% Au-60% Pd with 10% Rh-90% Pt.	Widely used.	73
Fuses			
Temperature-limiting fuses to protect elec. furnaces or their contents from overheating.	Au-Pd alloys.	These alloys have very narrow melting ranges. The Au-Pd alloy containing ~ 50% Au has max. hardness and resistivity.	17 66 73
Fuses in instrument and telecommunications work.	Ag-Pd alloys.		66
Detonation of explosives, ignition of rocket components, activation of various release mechanisms.	Pd-clad Al fuse wire (Pyrofuze). Al-Pd wt. ratio 3:1.	The exothermic reaction between the Pd case and the Al core goes to completion when elec. ignited. Pyrofuze functions in space as well as in the atmosphere.	73 76 82 193
Fuel Cells			
Electrodes, membranes, and catalysts.	Pd, e.g., wrought Pd anodes.	Use of Pd for fuel cells has not grown significantly since the use of fuel cells in space and military applications grows slowly.	34 67
Miscellaneous Electrical Uses			
Electron-tube grids.	Pd-clad Mo or W; Pd electroplate		69
Sensing element in gas analysis	Pd-Pt		69

TABLE A-XIII (Continued)

BRAZING ALLOYS

Use	Form of Palladium	Remarks	References
Use in gas-turbine, jet-engine, air-frame, missile, nuclear, and electronic fields to braze low-alloy and stainless steels; Ni, Co, and Cu-based alloys and refractory metals such as W and Mo. Use in joining ceramics to metals and in joining precious metals.	Binary, ternary, and quaternary alloys containing 5-60% Pd, such as Mn-Pd, Al-Pd, Ag-Pd, Ni-Pd, Al-Pd-Ag, Pd-Ni-Cu, Pd-Au-Cu, Ag-Cu-Pd, Ag-Pd-Mn, Ni-Mn-Pd, Pd-Ni-Li-B, and Pt-Pd-Au-Ag.	The graded m.p.'s of the alloys allow step brazing and simultaneous joining and heat-treating operations. Pd gives superior wetting and the alloys have little tendency to erode or dissolve the metals being joined. The probable max. service temperature of Pd and its alloys is 1200°C.	71 73
Joining the thin-walled Ni-alloy tubing in the F-1 liquid-propellant rocket engine that was used in the first stage of the Saturn V lunar space rocket.	20% Pd-75% Ag-5% Mn alloy.		76 192
Brazing steam-turbine blades of austenitic steels and Ni-base super alloys.	Various Pd-containing alloys.		71
Brazing stainless-steel honeycomb structure for stressed structural members subject to aerodynamic heating.	Ag-Pd-Mn alloy.	Alloys contain 20-33% Pd and are used for brazing at 1120-1220°C.	71
Constructing a stainless-steel fuel can for use in an "organic-cooled" nuclear reactor.	60% Pd-40% Ni alloy.	The alloy is used for brazing at 1230°C.	71
Brazing ceramic insulators to metal sheaths to protect thermocouples used in the nuclear field.	Pd-containing alloys, e.g. Ni-Pd-Mn and Ag-Cu-Pd.	Compositions of these two alloys available for brazing at 1120-1140°C and 700-1100°C, respectively.	71
Brazing Mo electrodes for electronically heated glass-melting furnaces.	Ag-Pd-Mn alloy.		71
Brazing W carbide tips to drill shanks for geological or mining operations.	Ag-Pd-Mn alloy.		71
Brazing in the assembly of internal components of large electronic valves.	Ag-Cu-Pd alloy.		71
Brazing radio-valve components.	Ag-Cu-Pd alloy.		71
Brazing electronic and other vacuum devices requiring vacuum-tight low-vapor-pressure seals.	40% Ni-60% Pd alloy or Au-Pd.		69

TABLE A-XIII (Continued)

BRAZING ALLOYS (Concluded)

Use	Form of Palladium	Remarks	References
Brazing Cu to itself or to other metals.	Ag-Cu-Pd alloys.	Cu brazing alloys are used in vacuum devices and for brazing components where Cd and Zn brazing alloys are unsuitable.	71
Brazing stainless steels used for high-temperature service, chiefly the austenitic Ni-Cr type.	Ni-Mn-Pd and Ag-Pd-Mn alloys.		71
Brazing gas-turbine blades of Ni and Ni alloys.	Ag-Pd-Mn and Ni-Mn-Pd alloys.		71
Brazing joints on low-expansion alloys based on Ni-Fe or Ni-Fe-Co, e.g. in glass-to-metal seals in vacuum-tube devices.	Pd-Ag-Cu alloy.	The low-expansion alloys are first Ni plated.	71
Brazing Mo to itself, to W, and to other metals.	60% Pd-40% Ni and Ag-Pd-Mn alloys.	Used commercially.	71
Brazing joints between Be and Monel or Cu-plated mild steel.	Ag-Cu-Pd	Joints are strong and vacuum tight.	71
Brazing Ti.	52% Ag-28$ Cu-20% Pd alloy.	This alloy gives some of the best strength figures for brazed Ti at ≤ 300°C.	71

GAS PURIFICATION

Use	Form of Palladium	Remarks	References
Diffusion units to produce ultrapure H_2, e.g. from cracked NH_3, for gas chromatography, for metallurgical atmospheres(e.g. for sintering and annealing refractory metals or processing high-Ni alloys), and for manufacturing semiconductor devices; to remove the H_2 carrier gas from the gas stream of gas chromatographs such as the apparatus of the Viking Lander Mars space probe; to admit H_2 into vacuum laboratory apparatus; to reduce the H_2 content of synthesis gas in preparation for chemical processing; and to separate and concentrate hydrogen isotopes.	Sponge Pd or 23-27% Ag-Pd alloys.	Pd hydrogen diffusers operate preferably at 450°C (850°F).	17 18 76 92 194 195 196 197 198
Recovery of olefins from gas mixtures or in gas-liquid chromatographic separation of olefins.	Pd(II) salts.	The olefins form complexes with Pd(II) that release the olefins on heating [German Patent 1,095,807 (1962)].	141

TABLE A-XIII (Continued)

Use	Form of Palladium	Remarks	References
LABORATORY AND PROCESS EQUIPMENT			
Liners for dies in glass-bottle manufacture.	Pd	Liners have shown no visible deterioration after feeding thousands of tons of low-fusing glass.	65
Spinnerets for producing synthetic fibers.	Au-Pd and Au-Pt-Pd alloys.		73
Laboratory crucibles.	Pd-Au alloy.		73
Cladding.	Ag-Au-Pd alloys.		69 73
Improvement of corrosion resistance of titanium and steel, e.g.			
Use in media fluctuating between oxidizing and reducing conditions such as petrochemical-plant tank linings and impellers for handling H_2SO_4.	Ti alloy containing 0.15-0.20% Pd.	A Ti alloy containing a small amount of Pd was used in the construction of a Wacker-process plant.	72 76
Equipment for handling chlorides.	Pd-Ti alloy.	The alloys resist crevice corrosion.	199
Increase passivity of stainless steel and Ti to oxidizing acids and of Ti to HCl.	0.1-1.0% Pd coatings.		17
Equipment for converting urea to melamine at 525-1100°F.			69
Improvement of heat transfer and condensation rates in condensers.	Surface coatings of thin Pd layers.		76
Trays for firing phosphors in oxidizing atmospheres.		See glass industry uses.	69
DENTAL USES			
Base for fired porcelain teeth.	Precision-cast Pd alloy.	The thermal expansion of Pt-group metals and their alloys falls in a range matching that of certain glasses and ceramics.	82
Supports in porcelain-overlay-type restorations.	Pd-rich alloys.		76

TABLE A-XIII (Continued)

DENTAL USES (Concluded)

Use	Form of Palladium	Remarks	References
Other specific dental alloys.	Low Rh-Pd alloys (usually with additional hardeners), Au-Cu-Pd.	Ag-Cu-Pd is a basis for many alloys.	73
Fired in porcelain teeth to give attachment to the precious-metal pin that is later soldered into place.	Ag-Au-Pd.		73
Toothpins and cast dental restorations to be covered with porcelain.	Au-Pt-Pd. Typical casting alloys: 90% Pd + Au-6-8% Pt-6% Ru or Rh and 86.5% Au-7.5% Pt-4% Pd- ~2% base metal.		73
Partial dentures and white alloys.	Alloys containing 3.5-18% Pd and ~3.4% Pt.		73
Early toothpin.	A thin Pd-Au sleeve fired onto porcelain (Au-clad Ni pin was soldered into the Pd-Au sleeve).		73
Current toothpin.	Specially processed Pd of U.S. Patent 2,766,527 (1956).	A thin layer of Pd between the Ni core and the Au cladding has come into worldwide use.	73
High-strength dental wires for orthodontic appliances and crown and bridge work.	Pd alloys with Pt, Au, Ag, Cu, Ni, and/or Zn plus fractional percentages of Ir, In, and Rh.	Contain 1-44% Pd. Solders do not contain Pd.	69
Cast Au alloys for inlays, crowns, bridges, and partial dentures.	Alloys of Au, Pt, Pd, Ag, Cu, and Zn.	Contain $\leq$ 0.5-20% Pd.	69

MEDICAL USES

Use	Form of Palladium	Remarks	References
Formerly in treatment of tuberculosis, gout, obesity, and rheumatoid arthritis.	Colloidal Pd compounds.		146 147
Research anticancer drugs.	Pd(II) complexes with 6-mercaptopurine and butylthiopurine.	Studied for leukopenic activity in chicks and/or activity against Adenosarcoma 755 and Sarcoma 180.	158 159

TABLE A-XIII (Continued)

Use	Form of Palladium	Remarks	References
MEDICAL USES (Concluded)			
Production of focal brain lesions.	Pelleted implant, e.g. of 78.7% Pd-21.3% Ni alloy.	The implant is heated inductively to produce the lesion [U.S. Patent 3,653,385 (1972)].	200
Therapeutic radiation sources.	Pd radioisotopes.	^{109}Pd complexes with hemato- and protoporphyrin have been recommended for lymphatic ablation in controlling homograft rejection.	157
JEWELRY AND DECORATIVE			
Jewelry, such as watchcases, cigarette cases, brooches, gem settings, earrings, and clips. Also used as a base coat before plating Au onto Ag.	For example: Most common alloy "white gold," containing > 15% Pd, 4.5% Ru, or 4% Ru + 1% Rh, as wrought and cast alloys; 50% Au-18% Ag-17% Pd-15% Cu (12-karat alloy); 5% Ni-Pd alloy; 4% Pd-96% Pt.	Pd is applied in thin flash deposits for plating watch and cigarette cases. Pd is generally plated onto a Ni base coat or Ag-plated Ni. The Pd-Pt alloy is used for difficult forming operations in Europe. U.S. stamping laws do not provide for Pd-Pt alloys.	69 73 76 201
Watch and clock bearings, springs, and balance wheels.	Pd alloys; Pd electrodeposited from baths containing $PdCl_2$.		194
Decorative on ceramics, glassware, etc.	Pd deposited by the "liquid-bright" method; unalloyed Pd leaf.		23 82
GLASS			
Containers and glass-working equipment in the manufacture of optical glass and other special-property glasses.	Pd-Rh alloy.	The wear-resistant alloy or Ir sometimes replaces Pt in the noncontaminating crucibles and tanks.	1 7
Trays for igniting the special fluorescent powders used in treating glass tubes for fluorescent lighting.	Pd metal?		65 73
Mirrors in astronomical instruments.	Pd metal?		194

TABLE A-XIII (Continued)

GLASS (Concluded)

Use	Form of Palladium	Remarks	References
Possibly used for the internal nucleation of glass crystallization.	Pd?	Approximately 0.001% Pt-group metal is added to the glass melt in the manufacture of ceramic-like materials made by heat-treating special glass compositions. Use of Pd in this application was not identified.	82
Metallized glass and ceramics.	Pd and alloys.	Metal film deposited by vacuum sublimation.	69
MISCELLANEOUS USES			
Electrolytic protection of steel-hull marine vessels.	20% Pd-80% Pt alloy or 50% Pd-50% Pt alloy.		69 73 84
Sensitization of plastics prior to electroless metal deposition.	Pd salt solutions.		C.A. indexes
High-temperature magnetic recording tape.	E.g.,electroplated intermediate barrier layers of Pd.		202
Magnetic memory devices.	E.g.,Pd deposited alternately with Co on a nonmagnetic support.	National Cash Register Co., French Patent 1,494,807 (1967).	203
Insecticide.	$(Ph_3P)_2CHPdBrCl_2$.	Patented by Monsanto Research Corp.	204
Photography			
Preparation of pictures to be transferred to porcelain and for toning solutions.	Pd compounds such as $PdCl_2$, K_2PdCl_4, and complex organic salts.		194
Image-receiving layers of film with and without Ag.	K_2PdCl_4, electrochemically precipitated Pd, $PdCl_2$, complex organic salts.	Probably noncommercial. Several recent patents by Kodak. Pd was formerly used in producing bluprints.	12 15 205-211
Patented anti-knock fuel additives.	Norbornadienepalladium(II) complexes.		141
Acceleration of W sintering.	Pd in small amounts.		82

TABLE A-XIII (Concluded)

MISCELLANEOUS USES (Concluded)

Use	Form of Palladium	Remarks	References
Separation of Cl_2 and I_2.	$Pd(NO_3)_2$.	The Merck Index was the only reference to mention this use.	194
CO alarm systems.	$PdCl_2$.	$PdCl_2$ easily reduced by CO.	76
Detecting CO to find leaks in buried gas pipes.	$PdCl_2$-impregnated paper.		194

TABLE A-XIV

CATALYTIC USES OF PALLADIUM

HYDROGENATION

Reaction	Form of Palladium	Remarks	References
Succinic anhydride $\longrightarrow$ butyrolactone	Pd/C or Pd/Al_2O_3	U.S. Patent 3,113,138. Not practiced com. Acids, anyhdrides, amides, or esters are resistant to hydrogenation over catalysts	73
Methylbutynol $\longrightarrow$ methylbutenol.	e.g., Pd/Al_2O_3, $Pd/CaCO_3$	Important in Vitamin A synthesis	76 C.A.* indexes
$HC{:}CH \longrightarrow H_2C{:}CH_2$ Disubstituted acetylenes $\longrightarrow$ cis-olefins	Pd/silica gel 50% $Pd/BaSO_4$ poisoned with $Pb(OAc)_2$ or quinoline	Acetylenes can be reduced to hydrocarbons or, if the amount of H_2 is limited or an inhibitor is added, to olefins. The corresponding olefin is formed most selectively and stereospecifically over Pd.	17 73 212 213
$RCOCl + H_2 \longrightarrow RCHO + HCl$	$Pd/BaSO_4$ + S-quinoline catalyst poison	Rosenmund reduction. Pd preferred in reductions of aldehydes to aromatic alcs. or hydrocarbons. Aliphatic aldehydes are reduced slowly. Pa useful for redn. of unsatd. to satd. aldehydes.	73 212
Purification of terephthalic acid by hydrogenation of impurities		Used com.	15
Redn. of aromatic systems, e.g., perhydrogenation of rosin at 5,000 psi and 200°C	$Pd/SrCO_3$	Preferred before advent of use of Rh and Ru to avoid hydrogenolysis of substituents. Pd causes less decarboxylation of rosin than other catalysts. Pd more effective catalyst as temp. and pressure are raised	73
Dehydrohalogenation (H replaces halogen)	Pd/C, $Pd/BaSO_4$	Pd most frequently used catalyst	73 212 C.A.* indexes
Hydrogenolysis of activated moieties such as allyl, vinyl, and benzyl groups	10% Pd/C, 5% $Pd/BaCO_3$	Pd preferred to Ni and Cu-Cr oxide in dibenzylations in which $PhCH_2$ is linked via O, N, or S. Most reported C-C bond cleavages over Pd catalysts.	73 212 214
Butyrolactone hydrogenation	Pd/C		47

*C.A. = Chemical Abstracts

TABLE A-XIV (Continued)

Reaction	Form of Palladium	Remarks	References
Aromatic ketones to aromatic alcs., e.g., 2-ethyl-, 2-*tert*-butyl-, or 2-chloro-anthraquinone to the corresponding anthraquinol	Pd/Al_2O_3	Used in com. H_2O_2 production: The anthraquinol is air oxidized to the quinone and H_2O_2. The quinone is hydrogenated in the presence of Pd, Pt, Ni or Raney-Ni at 25-40°C and 1-3 atm. Pd is the best known catalyst for selective hydrogenation of an aromatic ketone to an aromatic alc. and for hydrogenolysis of the latter to the aromatic hydrocarbon.	17 73 213 214 215
Aromatic nitro compds. to aromatic amines	Supported Pd	Proceeds at room temp. and pressure	213
Nitriles ⟶ primary, secondary, or tertiary amines	Pd/C	Formation of secondary and tertiary amines decreased by use of acidic media	73
Nitrocyclohexanone ⟶ cyclohexanone oxime or cyclohexylhydroxylamine	Pd/C	Widely studied reaction. Products are nylon intermediates.	73 C.A.* indexes
N-Nitrosodialkylamines ⟶ dialkylhydrazines	Pd/C	Product used for rocket fuel.	73 C.A.* indexes
Oximes ⟶ amines		Pd frequently used.	73
Hydrogenatinn of phenols and Ph ethers, e.g.,		Pd at high pressure formerly best catalyst.	73
Phenols ⟶ cyclohexanones	Pd/C		73, 214 C.A.* indexes
Reductive alkylations		Pt used 15 times more often than Pd. Pd used in condensations between active methylene compds. and alkylating agents under reducing conditions.	73
Vegetable oil hydrogenation		Widely studied. Usual com. catalyst for hydrogenating edible and nonedible oils is 25% Ni.	87 216 C.A.* indexes

*C.A. = Chemical Abstracts

TABLE A-XIV (Continued)

Reaction	Form of Palladium	Remarks	References
$NH_4NO_3 \xrightarrow[\text{buffer}]{\text{phosphate}} NH_2OH$ phosphate		The product is used to produce cyclohexanone oxime. Treating the oxime with oleum gives the nylon 6 intermediate caprolactam. The U.S. subsidiary of DSM will use this improved Stamicarbon-devised process in an Augusta, Georgia, plant of capacity 50,000 tons/year.	217
Hydrogenation of raw gas or tail end to remove acetylenes and diolefins from ethylene and propylene in steam-cracking plants	Pd/Al_2O_3	Catalysts produced by Catalysts and Chemicals, Dow, Girdler, and Engelhard. Catalysts do not wear out. Annual market based on initial charge of new plants is $250,000 (1972 prices).	87
Partial hydrogenation of pyrolysis gasoline, the steam-cracking co-product, to remove diolefins and most of the S or hydrogenation of the olefins to give olefin-free aromatic concs.	Pd/Al_2O_3	Total (new and replacement) market $100,000. Catalyst lifetime - 3 years.	87
	DEHYDROGENATIONS		
Aromatization, e.g., cyclic ketones $\longrightarrow$ phenols; Tetralin $\longrightarrow$ naphthalene	5-30% Pd/C	Most common dehydrogenation reaction catalyzed by Pd. Involved in petroleum reforming, org. synthesis, and org. structure elucidation, usually at high temp. or with a H acceptor. Batch-type processes are slow, requiring high catalyst-loading levels. Carriers other than C may allow regneration by oxidn. Fragmentation and isomerization may be involved in dehydrogenations of org. compds.	73, 76
	OXIDATIONS		
$CH_2{:}CH_2 + PdCl_2 + H_2O \longrightarrow MeCHO + Pd + 2HCl$; $Pd + 2CuCl_2 \longrightarrow PdCl_2 + 2CuCl$; $1/2\ O_2/air + 2HCl + 2\ CuCl \longrightarrow 2\ CuCl_2 + H_2O$ (Wacker process)	$PdCl_2$ and $CuCl_2$	One of the biggest identifiable uses of Pd in the chem. industry. There were 19 refs. to this or similar processes in 1962-1969; 12 were patents.	141, 218, 219
Similarly: Propylene $\longrightarrow$ acetone	$PdCl_2 + CuCl_2$		
1-Butene $\longrightarrow$ MeCOEt	$PdCl_2 + CuCl_2$		

TABLE A-XIV (Continued)

Reaction	Form of Palladium	Remarks	References
$CH{\vdots}CH \longrightarrow$ HCHO, MeCHO, EtCHO, HC:CCHO	$PdCl_2 + CuCl_2$		
$CH_2{:}CH_2 + PdCl_2 + AcOH \longrightarrow CH_2{:}CHOAc + Pd + 2HCl$	$PdCl_2 + CuCl_2$	A major com. process. In the period 1962-1968, there were 23 patents issued regarding the oxidn. of ethylene to vinyl acetate. Higher olefins give complex products due to isomerization.	140 141 218
Alcs. + $O_2 \longrightarrow$ ketones, aldehydes, or acids e.g., $HOC_6H_4CH_2OH \longrightarrow HOC_6H_4CHO$		Pt used more than Pd in catalytic oxidns. Pd is is superior in oxidn. of hydroxbenzyl alcs. to hydroxybenzaldehyde. French Patent 1,337,243 (1963). Pd useful in batch and continuous production of uronic acids by oxidizing protected aldoses (U.S. Patent 2,845,439).	73
Unsatd. aldehydes $\longrightarrow$ unsatd. carboxylic acids	$PdCl_2 + Cu(OAc)_2$ in AcOH		141
Aldehyde and an alc. $\longrightarrow$ ester	Pd salt	Japan. Patent 24,788 (1964).	73
$CH_2{:}CH_2 + H_2O + HNO_3 \longrightarrow$ CHOCHO, glyoxal	Pd	Japan. Patent 19,784 (1964).	73
$PhCH_2Cl$ or $PhCHCl_2$ + air + $NH_3 \longrightarrow PhCN$	10% Pd/C	German Patent 1,152,098 (1963).	73
$HCN \xrightarrow{300\text{-}800°C} N{\vdots}CC{\vdots}N$, cyanogen	PdO		97
$NH_3 \longrightarrow HNO_3$	90% Pt, 5% Rh, 5% Pd	Preferred catalyst 90% Pt-10% Rh.	37
	OXIDATIVE COUPLING		
Olefins or aromatic compds. $\longrightarrow$ coupled products, e.g.,	Pd(II) compds.		220
$ArCH{:}CH_2 + ArH \longrightarrow ArCH{:}CHAr$	$Pd(OAc)_2$		220
$C_6H_6 \longrightarrow C_6H_5{-}C_6H_5$	Pd(II) + NaOAc in AcOH	*cf.* French Patent 1,345,311 (1963).	73 220

TABLE A-XIV (Continued)

Reaction	Form of Palladium	Remarks	References
CARBONYLATION			
$CH_2:CH_2 + CO + PdCl_2 \longrightarrow ClCH_2CH_2COCl$, β-propionyl chloride, + Pd or $CH_2:CH_2 + CO + H_2O + PdCl_2 \longrightarrow CH_2:CHCO_2H$, acrylic acid + 2HCl + Pd	$PdCl_2 + CuCl_2$ in AcOH	α-Olefins incorporate CO exclusively in the terminal position. The reaction is patented for producing acid chlorides, carboxylic acids, esters, lactones and oxo acids.	140 141
$CH_2:CH_2 + CO + 1/2\ O_2 \longrightarrow CH_2:CHCO_2H + AcOCH_2CH_2CO_2H$ $AcOCH_2CH_2CO_2H \xrightarrow{\Delta} CH_2:CHCO_2H + AcOH$	$PdCl_2 + CuCl_2$ in AcOH		218
$CH_2:CH_2 + CO + H_2 \longrightarrow EtCHO + C_2H_6$	$PdCl_2 + CuCl_2$ in AcOH		220
$PdCl_2 + 2ROH + CO \longrightarrow Pd + ROCO_2R + 2HCl$	$PdCl_2$	U.S. Patent 3,153,094 (1964).	73
Carbonylation, hydrogenations, and transesterifications	Pd(II) salts bound to ion-exchange resins		221
POLYMERIZATION			
Olefin polymns., dimerizations, copolymns. with CO	Pd(II) complexes	U.S. Patents 3,194,800 (1965) and 3,330,815 (1967); Netherlands Patent 6,607,898 (1966) (U.S. Rubber).	141
e.g., $C_2H_4 \longrightarrow$ butenes	$[PdCl_2(C_2H_4)_2]_2$ + Cu(II) or Fe(II) salts		141
Butadiene $\longrightarrow$ 1,2-polybutadiene	$PdCl_2$		141
Butadiene $\longrightarrow$ _trans_-1,4-polybutadiene, (very low yield)	PdI_2 or $Pd(CN)_2$		141
$HC\vdots CH \longrightarrow$ _trans_-polyacetylene, benzene, other by-products	$PdCl_2$	Acetylene and monosubstituted acetylenes form labile Pd complexes that polymerize readily.	141
Disubstituted acetylenes $\longrightarrow$ cyclobutadienes	Pd(II)	Widely studied reaction.	141
$HC\vdots CH \longrightarrow$ polyacetylene	10% Pd/C, absence of H_2		73

TABLE A-XIV (Continued)

Reaction	Form of Palladium	Remarks	References
Synthesis of phenoxyoctadiene, octatriene, and alkylated phenols; telomerization of, e.g., butadiene	π-allylpalladium complexes		222
Allenes ⟶ butadienes	$Pd(OAc)_2$ or $Pd(NO_3)_2$ in AcOH		222
Allenes ⟶ polymers contg. conjugated unsatn.	$Pd(OAc)_2 + Ar_3P$		222
Ketenes ⟶ solid polymers of M.W. ≤ 2,000	$PdCl_2$-PhCN	Belgian Patent 638,289 (1964).	73
ISOMERIZATION			
Olefin isomerizations	$PdCl_2$, $[PdCl_2(PhCN)_2]$, Pd(II)-olefin and π-allyl complexes, Pd(II)-phosphine complexes, bis(acetylacetonato)palladium(II), and Pd(O) complexes	No patents issued in 1962-1968 regarding this use	141
Cis-trans isomerizations, e.g., in studies of equilibria between satd. *cis*-and *trans*-isomers.		Occur during hydrogenations	73 76
Double bond migrations, e.g., an exo- to an endomethylene double bond in the absence of H_2	Pd/C, $Pd/CaCO_3$	Pd the most active of the Pt-group metals. Probably occurs during most olefin hydrogenations over Pd.	73 214
Skeletal isomerizations, e.g., of bicycloheptanones			73
Functional-group migrations			73
FRAGMENTATION			
Decarbonylation			
PhCH:CHCHO ⟶ $PhCH:CH_2$ + CO	10% Pd/C	Pd exceptionally active	73
Furfural ⟶ furan + CO (162-230°C)		Very long catalyst life in presence of added base. U.S. Patnet 3,007,941 (1961).	73
Decarboxamidation			
β-ketoamides $\xrightarrow{\Delta}$ Ketone + NH_3 + CO	Pd		73

TABLE A-XIV (Concluded)

Reaction	Form of Palladium	Remarks	References
Deamination			
$PhCH_2NH_2 \xrightarrow[\text{reflux}]{} (PhCH_2)_2NH + NH_3$	$Pd/BaSO_4$		73
	DISPROPORTIONATION		
 3 cyclohexene → benzene + 2 cyclohexane	Pd/C, $Pd/CaCO_3$	Disproportionation is used com. to stabilize natural products such as abietic acid. Frequently less than theoretical amt. of Pd is used in hydrogenation to avoid formation of difficult-to-reduce aromatic systems.	73 213 214
	HYDROSILATION (Silylation)		
Olefin + silane ⟶ alkylsilane	PdI_2, $PdCl_2$, PdO, Pd-phosphine complexes	Pt complexes are used com.	141 C.A.* indexes
	HYDROGEN-TRANSFER REACTIONS		
Cyclodec-1-en-6-ol ⟶ cyclodecanone: 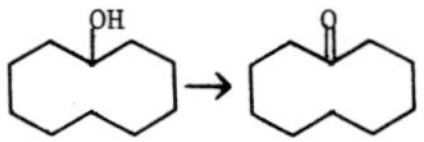		Pd is the preferred catalyst for these reactions. Pd catalyzes H transfer from a H donor such as cyclohexene to H acceptors in the absence of mol. H_2.	73 213

*C.A. = Chemical Abstracts

TABLE A-XV

1975 CATALYTIC CONVERTER SYSTEMS[37,101,109]/

	General Motors Underfloor	Ford	Chrysler
Amount of platinum in catalyst, troy oz	~0.035-0.036	0.07-0.10	< 0.1
Amount of palladium in catalyst, troy oz	~0.014-0.15	0.006-0.05	< 0.1
Substrate	Pellets	Monolith	Monolith
Number per car	1	2 (V-8)* 1 (4,6)	1
Location	Underfloor	Toeboard	Underfloor

* One converter on Maverick, Comet, Econoline, and Bronco.

TABLE A-XVI

INCO SMELTER STACK EMISSIONS, 1971[223/]

	Dust Loss (tons/year)	Troy oz (Pd/ton)	Annual amount of Pd lost (oz)
Orford stack	2,290	0.283	648
Converter roof monitors	1,697	-	-
Nickel stack	34,054	0.033	1,124
Copper stack	2,163	0.141	305
Coniston sinter stack	2,546	0.065	166
Coniston smelter stack	2,774	0.065	178
Iron ore stack	1,972	0.001	2
No. 1 pellet sinter machine	872	0.009	8
No. 2 pellet sinter machine	767	0.009	7

APPENDIX B

PHYSICAL PROPERTIES OF PALLADIUM

Palladium has the lowest specific gravity of the platinum metals, the lowest melting point, the highest coefficient of linear expansion, the highest electrical resistivity (only slightly higher than that of platinum), and the lowest temperature coefficient of resistance.[66]

Table B-I gives some of the more important physical properties of palladium and Table B-II gives physical properties of the more common palladium compounds.

Properties of binary and ternary alloys, electrical contact materials, and brazing alloys can be found in references 69, 71, and 76.

Thermodynamic data and oxidation potentials for numerous inorganic palladium compounds and complexes have been compiled by R. N. Goldberg and L. G. Hepler.[224] The heats of formation at 298°C of the crystalline forms are given in Table B-II.

TABLE B-I PHYSICAL PROPERTIES OF PALLADIUM AND SOME OF ITS ALLOYS

Property	Pure Palladium	45% Ru-Pd	40% Ag-Pd	40% Cu-Pd		References
				Quenched	Ordered	
Vapor pressure, torr,						
1500°C	10^{-2}					225
m.p. (1552°C)	2.6×10^{-2}					72
Atomic magnetic susceptibility, χ^A ($cm^2/g/x10^6$	+558.1					85
Specific gravity, 25°C g/cm^3	12.02 (0.43 lb/in^3)	12.0	11.3	10.6		72
Linear coefficient of thermal expansion at 20°C, per °C	11.6×10^{-6}					72
Specific heat at 0°C, cal/g	0.0584					2
Thermal conductivity	0.18					2, 72
at 20°C, cal/(sec) (cm^2)(°C/cm)	0.17					17, 85, 226
Electrical resistivity μΩ-cm						
0°C	9.93	22.9	42.0	34.8	3.3	2, 72
20°C	10.8	23.5	42.0	35.0	3.5	72, 226
Mass susceptibility, cgs units	5.23×10^{-6}					72

TABLE B-I (Concluded)

Property	Pure Palladium	45% Ru-Pd	40% Ag-Pd	40% Cu-Pd		References
				Quenched	Ordered	
Work function	4.99					72
Tensile strength, annealed,psi	24,000-30,000					72
Young's modulus of elasticity, psi	17×10^6					72
Elongation %	24-30 40 (annealed)	25	47			72 66
Vickers hardness	50 37-39 (annealed)	85	100			72
Temperature coefficient of resistivity, 0-100°C	0.0038	0.0013	0.00002	0.00032	0.00224	72, 76
First ionization potential, ev	8.33					30
Second ionization potential	19.42					30
Ionic potential	1.67					227
Electronegativity	1.35					227
Stability constant (log K)	18.5					227
Reflectivity at 7500 Å	66%					23

TABLE B-II

PHYSICAL PROPERTIES OF PALLADIUM COMPOUNDS

	Oxidation State	Boiling Point (°C)	Melting Point (°C)	Solubility	Physical Form	ΔH_c, kcal/Mole	Remarks	References
$(NH_4)_2PdCl_6$ Ammonium hexachloropalladate(II)	IV			Slightly sol. in H_2O.	Red cryst. powder, octahedral			72 228
Na_2PdCl_6	IV						Prepd. by treating Pd with hot aqua regia and NaCl. Analogous salts prepd. similarly.	224
K_2PdCl_6	IV		Evolves Cl_2 at >175°	Insol. in alc. or in H_2O contg. alkali-chlorides; sol. in H_2O and hot dil. HCl.	Red or brown-red crystals	~ -277	Reagent for CO detn.	194 224 229 230
$Pd[PdF_6]$ Palladium(II) hexafluoropalladate(IV)	II, IV	Decomp.	Decomp.	Decomp. in H_2O; sol. in aq. HF	Black, hygroscopic rhombic crystals		Decomp. on contact with H_2O with O_2 liberation. A strong oxidizing agent. Formed by direct combination of the elements at high pressure or by fluorination of PdX_2 (X=halogen) at 200-250°C.	224 226 230 231 232
$PdO_2 \cdot xH_2O$	IV		Decomp. with loss of H_2O and O_2.	Sol. in dil. acid and concd. alkali solns. Insol. in cold or hot H_2O	Dull red		Loses O_2 slowly at room temp.; transformed to PdO completely at 200°C. A strong oxidizing agent. Prepd. by treating $PdCl_6^{-2}$ with alkali. Existence of anhyd. PdO_2 and higher oxides not firmly established.	224 228 230 231 233
$Pd_2O_3 \cdot xH_2O$ Palladium sesquioxide	II, IV			Sol. in HCl.	Brown powder		Decomp. violently on heating. Prepd. by anodic oxidn. of $Pd(NO_3)_2$ at 0°.	72 228

TABLE B-II (Continued)

PHYSICAL PROPERTIES OF PALLADIUM COMPOUNDS

	Oxidation State	Boiling Point (°C)	Melting Point (°C)	Solubility	Physical Form	ΔH_c, kcal/Mole	Remarks	References
PdS_2	IV		Decomp.	Insol. in strong acids, hot H_2O. Sol. in aq. regia, CS_2, and $(NH_4)_2S$.	Chocolate-colored solid.	~-19	Dissociates at >600° to PdS. Prepd. by heating PdS, $PdCl_2$, K_2PdCl_6, or Rb_2PdBr_6 with S at 200-500°C or by treating Na_2PdS_3 with acid.	72 213 224 230 234
PdSi	IV			Insol. in cold H_2O	Cryst.			231
$Pd(CN)_4$	IV			Insol. in H_2O	Pale rose		Less stable than $Pd(CN)_2$; slowly decomp. to give HCN	72
$K_2Pd(CN)_6$	IV						Stable to water and dil. acids at room temp. Slowly decomp. to $K_2Pd(CN)_4$ and $Pd(CN)_2$ in boiling water. Prepd. by treating K_2PdCl_6 with KCN in the presence of a strong oxidizing agent	226
$(NH_4)_2PdCl_4$	II			Sol. in H_2O, insol. in alc., Et_2O	Yellow-green prisms and needles			226 228
Na_2PdCl_4	II			Sol. in H_2O, alc.; slightly sol. in acetone, AcOH, EtOAc	Red deliquescent prisms		$PdCl_2 + NaCl \xrightarrow{HCl} Na_2PdCl_4$	226 228
$Na_2PdCl_4 \cdot 3H_2O$	II			Sol. in alc. and H_2O	Brown hygroscopic salt		Used in analysis (testing for CO, C_2H_4, illuminating gas, I_2). Reduced by alk. HCHO in the presence of a carrier to prep. a catalyst.	47 229
K_2PdCl_4	II			Sol. in H_2O, aq. alc.	Red-brown crystals	-259	Evolves HCl at 105°. Used in photography	194 224 228

TABLE B-II (Continued)

PHYSICAL PROPERTIES OF PALLADIUM COMPOUNDS

	Oxidation State	Boiling Point (°C)	Melting Point (°C)	Solubility	Physical Form	ΔH_c, kcal/Mole	Remarks	References
K_2PdBr_4	II				Red-brown needles		Air stable. Prepd. by halogen exchange with K_2PdCl_4.	226 228
$Na_2Pd(NO_2)_4$	II			Sol. in H_2O			The basis for a useful plating bath. Gives moderately thick bright deposits.	72 73
PdF_2	II	Decomp. red heat (impure)	Volatilizes (impure)	Slightly sol. in cold H_2O, sol. in HF (impure)	Pale violet, rutile structure		Prepd. by treating $Pd(NO_3)_2$ with HF. Hydrolyzed in H_2O to the hydroxide.	224 226 231 232
$PdCl_2$	II		Decomp. 500°	Sol. in hot and cold H_2O, HCl, alc., acetone, alkali metal chloride solns.	Cubic needles, dark red crystals, deliquescent orange-red powder	~-39	Formed from the elements at 500°C. Moist $PdCl_2$ is reduced by CO at $\geq$250°. Used in CO detection, plating baths, starting material for other Pd compds. Dissoc. to Pd and Cl_2 at $\geq$600°. Cl_2 evolved at 160° in air.	72 224 228 229 231 232 234
$PdCl_2.2H_2O$	II		Decomp.	Sol. in H_2O, alc., HCl, and acetone	Red-brown prisms, deliquescent			194 228 231
$Pd(NH_3)_2Cl_2$	II		Decomp.	Sol. in aq. NH_3 to give $Pd(NH_3)_4Cl_2.H_2O$. Sol. in hot, concd. HCl; H_2SO_3; potash lye; hot H_2O with decompn.	Yellow tetragonal prisms	~-101	Precursor of Pd sponge in refining operation. Used in electroless Plating and bright Pd plating.	72 73 224 226 228 235

TABLE B-II (Continued

PHYSICAL PROPERTIES OF PALLADIUM COMPOUNDS

	Oxidation State	Boiling Point (°C)	Melting Point (°C)	Solubility	Physical Form	ΔH_c, kcal/Mole	Remarks	References
$Pd(NH_3)_4Cl_2.H_2O$	II		Decomp. 120°C	Sol. in H_2O	White tetragonal crystals	~-154 (anhyd.)	Converted to insol. $Pd(NH_3)_2Cl_2$ by adding HCl. Loses NH_3 and H_2O at 120° to give $Pd(NH_3)_2Cl_2$.	73 224 228 235
[$Pd(PPh_3)_2Cl_2$] Dichlorobis(triphenyl-phosphine)	II				Yellow		Air stable. Used in basic research. Numerous catalyst uses.	228 237 238
$PdBr_2$	II		Decomp.	Insol. in H_2O and alc.; sol. in HBr.	Red-brown mass; not characterized structurally	~-25	Decomp. in H_2O. Prepd. by treating Pd with Br_2 and HNO_3.	224 228 231 232
PdI_2	II		Decomp. 350°	Insol. in H_2O, HCl, dil. H_2SO_4, alc., ether. Slightly sol. in excess KI, H_2SO_3, NH_3 (with decompn.), hot concd. HNO_3.	Dark red to black powder. Not characterized structurally.	-14	Dissocn. pressure 1 atm. at 235°. Begins evolving I_2 at 100°C. Pptd. from $PdCl_2$ solns. by adding I^-.	224 228 232
PdO	II		Decomp. >550°; m. 870°	Insol. in H_2O, acids; slightly sol. in aqua regia; sol. in 48% HBr.	Black powder or black-green or amber mass	-27.6 [83(gas)]	Formed by heating Pd in air at red heat. PdO catalyst is prepd. by fusing $PdCl_2$ and $NaNO_3$ at ~600°C. Strong oxidizing agent. Hydrogenation catalyst.	47 194 224 226 228 231 234 235 239
$PdO.\underline{x}H_2O$	II		Decomp.	Insol. in H_2O; sol. acids, NH_3, NH_4Cl	Yellow to brown			235

TABLE B-II (Continued)

PHYSICAL PROPERTIES OF PALLADIUM COMPOUNDS

	Oxidation State	Boiling Point (°C)	Melting Point (°C)	Solubility	Physical Form	ΔH_c, kcal/Mole	Remarks	References
$Pd(OH)_2$	II		Decomp.	Sol. in acids to give Pd(II) salts and alkalies to give palladites (PdO_2^{2-}).	Yellow-brown to black	-88	Decomp. to Pd on heating to ~800°. Used in the prepn. of catalysts by redn. with HCHO, HCO_2H, or H_2. Soly. product 1.1×10^{-29}. Stable in H_2O.	72 224 226 228 240
$Pd(O_2CCH_3)_2$	II			Decomp. in H_2O. Sol. in glacial AcOH, hot benzene, org. amines, HCl (with decompn.), $CHCl_3$, CH_3CN.	Yellow-tan crystals		Used as a catalyst for liq.-phase oxidn. of hydrocarbons and liq.-phase reactions in org. solvents. Forms adducts with benzene and org. amines. Hydrolyzes slowly in cold H_2O, rapidly >85°. Decomp. in air at >85° to Pd.	72 226
$Pd(O_2C_5H_7)$ Bis(2,4-pentanedionato)-palladium(II) [Palladium(II) acetylacetonate]	II			Sol. in hot benzene and $CHCl_3$	Yellow crystals		Isomerization catalyst. Other research catalytic uses	15 141 220 226
PdS	II		Decomp. 950°	Insol. in cold and hot H_2O, dil. HCl, $(NH_4)_2S$; sol. in HNO_3, aq. regia	Brown-black or blue powder depending on prepn.	~-18	Converted slowly into a basic sulfate when heated in air. At higher temps., Pd forms. Prepd. from the elements or by treating a Pd(II) salt with H_2S.	72 224 226 228 230 231

TABLE B-II (Continued)

PHYSICAL PROPERTIES OF PALLADIUM COMPOUNDS

	Oxidation State	Boiling Point (°C)	Melting Point (°C)	Solubility	Physical Form	ΔH_c, kcal/Mole	Remarks	References
$PdSO_4 \cdot 2H_2O$	II		Decomp.	Very sol. in cold H_2O, decomp. in hot H_2O	Red-brown deliquescent crystals		Hydrated PdO ppts. from aq. solns. $PdSO_4$ with ammonium molybdate on silica gel is used in gas-detector tubes for CO.	72, 230, 231
$PdSeO_4$			Decomp. at red heat	Very sol. in cold H_2O and NH_3; insol. in alc., ether, alkali	Dark brown-red rhombic, deliquescent			235
PdSe	II		960°	Insol. in cold H_2O. Sol. in aq. regia	Dark gray or brown solid		Prepd. from $PdCl_2$ and H_2Se or Pd and Se	226, 231, 235
$[Pd(NH_3)_2(NO_2)_2]$	II			Sparingly sol. in cold H_2O, more sol. in hot H_2O, sol. in aq. NH_3	Yellow-white crystals		Adding to aq. NH_3 gives the salt complex $[Pd(NH_3)_4](NO_2)_2$, which is used for electro-depositing Pd ("P"-salt bath)	72
$Pd(NH_3)_4(NO_2)_2$	II						Used com. for plating	73
$Pd(NO_3)_2$	II		Decomp.	Sol. in dil. HNO_3; sol. in H_2O with trubidity.	Brown, rhombic, deliquescent crystals		Hydrated PdO ppts. from aq. solns. as a result of hydrolysis. Prepd. from Pd and hot concd. HNO_3 Anal. reagent.	194, 224, 229, 231
$[Pd(NH_3)_2(NO_3)_2]$	II				Brown solid		Detonates violently when heated. Used in electro-plating.	14, 228
$[Pd(NH_3)_4](NO_3)_2$	II			Decomp. by HCl. Sol. in H_2O, HNO_3, NH_3; insol. in alc.	Transparent prisms and plates		Detonates slightly after melting.	228

TABLE B-II (Concluded)

PHYSICAL PROPERTIES OF PALLADIUM COMPOUNDS

	Oxidation State	Boiling Point (°C)	Melting Point (°C)	Solubility	Physical Form	ΔH_c, kcal/Mole	Remarks	References
$Pd(CN)_2$	II		Decomp. 210°	Insol. in cold and hot H_2O and dil. acid; sol in KCN, NH_4OH, HCl.	Light yellow gelatinous ppt.		Prepd. by treating a Pd(II) salt with $Hg(CN)_2$ or warming $K_2Pd(CN)_4$ with HCl. Forms adducts with NH_3 and org. bases.	72 226 228 231
Pd	0	2900 [92] 2927 [235] 3980 [96]	1552	Sol. in HNO_3, hot concd. H_2SO_4, aq. regia, HCl-$HClO_3$ mixt. Slightly sol. in concd. HCl. Sol. in fused alkalis.	Silver-white metal, octahedral cubic structure; black powder; spongy mass.	0 [91(gas)]	Appreciably volatile at high temps. Converted to PdO at red heat. Oxide coating that forms at 600° decomp. at 800°.	2 92 96 194 229 231 235
$Pd(PPh_3)_4$	0		Decomp. 100-105°C		Yellow solid		Air sensitive. Used in basic chem. research (academic or industrial), e.g. to prepare complexes having palladium-carbon-sigma bonds. Numerous catalyst uses.	141 237 241 C.A.* index

*C.A. = Chemical Abstracts

APPENDIX C

CHEMISTRY

Palladium in its chemical compounds is quadrivalent, generally divalent, and rarely univalent. Reducing agents such as carbon monoxide or hydrogen readily reduce many palladium salts to the metal. Organic compounds and less noble metals reduce palladium salts in aqueous solutions.[76]

Table B-II (Appendix B, p. 98) includes the physical and chemical properties of some of the common and more generally useful palladium compounds.[76]

Palladium complexes are important in analysis, refining, and electroplating. Palladium(II) complexes such as PdX_4^{-2} where X=Cl, Br, CN, NO_2, and SCN and PdL_4^{+2} where L=various amines exist in aqueous media. Palladium(0) complexes include $Pd(CNR)_2$ where R=phenyl, etc.; $Pd(diars)_2$; $Pd(PAr_3)_4$; and $Pd(PF_3)_4$.[76]

A few univalent palladium complexes have been reported. These include $[PdAl_2Cl_7(C_6H_6)]_2$, $[Pd(C_6H_6)(H_2O)ClO_4]_n$, $[PdCl(\underline{tert}\text{-}BuNC)_2]_2(C_6H_5Cl)$, $[PdBr(\underline{tert}\text{-}BuNC)_2]_2$, $[PdI(\underline{tert}\text{-}BuNC)_2]_2$, $[PdI(PPh_3)(\underline{tert}\text{-}BuNC)]_2$, and $[PdI(Ph_2PCH_2CH_2PPh_2)]$.[226,242]

Following are brief discussions of the corrosion resistance of palladium, its oxidation and volatilization, its organic chemistry, palladium carbonyls, and palladium alloys.

A. Corrosion

Although all of the platinum-group metals--ruthenium, rhodium, palladium, osmium, iridium and platinum--are chemically noble, they differ significantly among themselves in reactivity. Tables C-I and C-II compare the corrosion behavior of palladium with that of platinum, for which, because of its lower price, palladium is often substituted.

Palladium is not attacked at room temperature by sulfuric, hydrochloric, hydrofluoric, acetic, and oxalic acids.

Strongly oxidizing substances such as nitric acid, hot sulfuric acid, ferric chloride, hypochlorite, chlorine, and bromine attack palladium. Palladium is corroded anodically in hydrochloric acid or acidic chloride solutions.[76] Precipitated palladium is corroded quantitatively by hydrochloric acid.[243] Although molten sodium peroxide, hydroxide, and carbonate a[illegible] group IA and IIA oxides attack palladium, molten sodium nitrate does not.[76]

TABLE C-I

CORROSION RESISTANCE OF PALLADIUM AND PLATINUM[84,197]

Corrosive Agent, Temperature	No Appreciable Corrosion		Some Attack But Not Enough to Preclude Use		Attacked Enough to Preclude Use		Rapid Attack	
	Pd	Pt	Pd	Pt	Pd	Pt	Pd	Pt
Concd. H_2SO_4 at 100°C		X			X			
Concd. H_2SO_4 at R.T.*	X	X						
H_2SeO_4,sp.gr.1.4, at 100°C						X	X	
H_2SeO_4, sp.gr.1.4, at R.T.	X	X						
H_3PO_4 at 100°C		X	X					
$HClO_4$ at 100°C		X			X			
$HClO_4$ at R.T.	X	X						
95% HNO_3 at 100°C		X					X	
95% HNO_3 at R.T.		X					X	
70% HNO_3 at R.T.		X					X	
2 N HNO_3 at R.T.		X			X			
1 N HNO_3 at R.T.		X	X					
0.1 N HNO_3 at R.T.	X	X						
Aqua regia at R.T.							X	X
40% HF at R.T.	X	X						
36% HCl at 100°			X	X				
36% HCl at R.T.	X	X						
HBr, sp.gr. 1.7, at 100°C							X	X
HBr, sp.gr. 1.7, at R.T.		X					X	
HI, sp.gr. 1.7, at 100°C							X	X
HI, sp.gr. 1.7, at R.T.		X					X	
Glacial acetic acid at 100°C	X	X						
Moist or dry liquid bromine at R.T.						X	X	
Moist chlorine at R.T.				X			X	
Dry chlorine at R.T.				X	X			
Moist or alc. iodine at R.T.		X	X					
Dry iodine at R.T.	X	X						
Bromine water at R.T.		X	X					
Moist H_2S at R.T.	X	X						
NaOCl soln. at 100°C		X					X	
NaOCl soln. at R.T.		X			X			
KCN soln. at 100°C						X	X	
KCN soln. at R.T.		X			X			
10% $FeCl_3$ soln. at 100°C							X	
10% $FeCl_3$ soln. at R.T.					X			
$HgCl_2$ soln. at 100°C	X	X						
$CuCl_2$ soln. at 100°C		X	X					
$CuSO_4$ soln. at 100°C	X	X						
$Al_2(SO_4)_3$ soln. at 100°C	X	X						

*R.T.-Room Temperature

TABLE C-II

CORROSION IN FUSED SALTS[197]

(Wt. loss in mg/dm^2/day, 1-hr exposure)

Salt	Temp., °C	Platinum	Palladium
$KHSO_4$	440	72.0	432.0
KCN	700	28,000.0	32,000.0
NaCN	700	7,450.0	14,200.0
1 KCN + 2NaCN	550	840.0	8,160.0
NaOH, reducing conditions	350	--	192
Na_2O_2	350	0.0	360
Na_2CO_3, reducing conditions	920	72.0	108

Fusions with potassium and sodium hydroxides give black porous coatings on palladium. Palladium is attacked somewhat by fused potassium hydrogen sulfate.[243]

Red-hot palladium is attacked by phosphorus, arsenic, silicon, sulfur, selenium, tellurium, and carbon.[17] The affinity of palladium and platinum for silicon can lead to disastrous consequences when they are heated under reducing conditions in contact with siliceous materials.[244] Palladium is tarnished by hot gases containing sulfides.[17] Hydrogen sulfide at >600°C (1110°F) attacks palladium to give a low-melting phase.[76] Cyanide solutions with an oxidizing agent are used for metallographic etching of palladium.[76]

Palladium, when heated in sulfur dioxide for one hour at 800-1000°C, forms a purple-blue film and gains weight but is not embrittled. Palladium catalyzes the oxidation of sulfur dioxide to sulfur trioxide in outdoor industrial exposures. This catalytic effect may be involved in the formation of a basic sulfate on the metal surface, which causes it to discolor.[197]

Palladium-silver jewelry and electrical contact alloys containing ≥50% palladium are attacked by nitric acid.[197] In household atmospheres, slightly more than 50% palladium prevents silver alloy tarnishing.

B. Oxidation and Volatilization

Palladium powder or massive palladium in air or oxygen at 350-790°C (660-1450°F) gains weight by forming a surface film of palladium monoxide.[197] Palladium monoxide film formation occurs even at 40 microns oxygen pressure at 390 and 450°C.[239] Above 790°C[69,197], (550°C[239]), palladium monoxide decomposes to the elements, its dissociation being complete at 870°.[243]

Before palladium monoxide film formation, a three-dimensional surface compound forms when palladium is heated at 180-300°C in air.[245]

Above 1000°C, oxygen dissolves in palladium and acts to increase its weight; however, at these temperatures, loss of palladium by volatilization tends to decrease its weight.[76] For example, at 1200°C in oxygen, palladium sheet gains ~0.1% in weight in the first hour by absorbing oxygen without forming a superficial oxide film. Heating for more than one hour at 1200°C gives a slow weight loss due to volatilization. On cooling, the sheet retains oxygen in solid solution.[243] The vapor pressure of palladium at 1500°C is about 0.01 mm compared with 0.000001 mm for platinum.[225] Between 1200 and 1700°K, the vapor pressure, P, of palladium is described by the equation

$$\text{Log } P_{mm} = 8.749 - \frac{18,655}{T}$$ [132]

where T = the temperature in °K. Below 1400°C, palladium in air volatilizes at a rate four orders less than in vacuo.[246]

At 1400, 1375, and 1325-50°C, palladium loses 4.8×10^{-9}, 2.0×10^{-9}, and 0.3×10^{-9} g/cm^2 sec, respectively. The equilibrium constant for palladium monoxide gas formation at 1400°C is 3,800 times that for palladium volatilization. The greater rate loss beginning at 1375°C is ascribed to formation of unstable palladium monoxide gas, which supplements the loss of palladium by volatilization.[246]

The rate loss of palladium to the vapor phase both as metal and gaseous oxide in industry at high temperatures can be reduced by alloying.[247]

C. Organic Chemistry

Several recent publications consider the chemical behavior of palladium and its compounds, especially organic complexes.

The preparation and properties of olefin and acetylene complexes of platinum and palladium were discussed by F. R. Hartley in an extensive review[141] with 605 references covering the literature from 1962 to 1968. A more recent work by Hartley, The Chemistry of Platinum and Palladium,[226] 1973, is devoted mainly to both organic and inorganic complexes of these elements. The book gives methods of preparing the most useful complexes.

The Organic Chemistry of Palladium[248] by P. M. Maitlis, published in 1971, discusses monoolefin, acetylene, diene, π-allylic, cyclopentadienyl, and benzene complexes as well as palladium hydride and compounds having palladium-carbon σ bonds and reactions catalyzed by palladium complexes.

Much information in this report on the use of palladium hydrogenation catalysts was obtained from the 1967 publication Catalytic Hydrogenation over Platinum Metals by P. N. Rylander[47] and the chapter by Rylander in E. M. Wise's Palladium. Recovery, Properties, and Uses, 1968.[73]

Palladium, usually in the +2 oxidation state, forms with unsaturated organic compounds a great variety of complexes that are sufficiently reactive to be intermediates in catalytic processes.[248] Olefin-palladium complexes are intermediates in the oxidation of ethylene to acetaldehyde (Wacker Process) and to vinyl acetate.[249]

The intermediate in the Wacker Process that decomposes in water to give acetaldehyde is only stable under an ethylene atmosphere.[141] This is because ethylene and other olefins coordinated to palladium(II) are readily attacked by nucleophiles.[249] Treating complexes of palladium and ethylene or another olefin with alcohols, primary amines, amides, chloride, or cyanide gives vinyl ethers or acetals, secondary amines, N-alkenylamides, vinyl chloride and vinyl cyanide, respectively. In all these reactions, palladium(II) is reduced to palladium(0) and must be reoxidized by, e.g., cupric salts or benzoquinone.[141]

Other homogeneous processes involving palladium-olefin complexes are carbonylation, hydrogenation, isomerization, and polymerization.[249] Carbonylation reactions using palladium-based catalysts have been reviewed by L. Cassar, et al.[250]

Olefins coordinated to palladium(II) can be arylated with aromatic compounds such as benzene, ferrocene, and heterocyclic aromatic compounds. For example, heating the styrene-palladium(II) chloride or acetate complex in acetic acid gives trans-stilbene and α-phenylethyl acetate. Acetate ion is essential to the reaction. Again, palladium(0) is produced in the reaction and must be reoxidized to palladium(II). Aromatic substitution of olefins involving palladium salts has been reviewed by I. Moritani and Y. Fujiwara.[251]

In nonaqueous solvents, lower olefins give π-olefin complexes, $[PdCl_2(olefin)]_2$, that are stable for a few hours in air.[141]

Treating olefins that have allylic hydrogen atoms with palladium salts in the presence of a weak base gives π-allylpalladim complexes. The more highly branched the olefin, the more π-allyl complex formed. Palladium-π-allyl complexes are yellow or red, air stable, and hydrolyze at room temperature. Thermal decomposition of these complexes gives dienes, whereas further oxidation gives α,β-unsaturated carbonyl compounds.[226,252]

A new alkylation method involves activation of a carbon atom α to a C-C double bond by formation of a π-allylpalladium(II) complex susceptible to nucleophilic attack by a polarizable carbanion in the presence of $\geq$ 4 equivalents triphenylphosphine.[253]

Treating tetrakis(triphenylphosphine)palladium(0) with chloroolefins gives olefin compexes with palladium-carbon σ bonds. However, olefins such as maleic anhydride, ethyl fumarate, and tetracyanoethylene give normal olefin complexes of palladium(0). Palladium(0)-monoolefin complexes are white crystalline solids stable to air oxidation but labile to ligand exchange.[141]

Olefins coordinated to palladium(II) form σ bonds by donation of π electrons from the olefin to vacant palladium orbitals. Electrons from a filled palladium *d* orbital can also be back-donated into vacant anti-bonding orbitals of the olefin. In palladium(0) complexes, the π character of the bond is almost entirely absent. Nonconjugated dienes form stronger complexes due to formation of two olefin-metal bonds.[220,249]

Palladium(0) complexes of acetylene involve strong π-back-donation from palladium, which renders acetylene susceptible to electrophilic attack. In palladium(II)-acetylene complexes, σ-donation of charge from acetylene to palladium is more important.[254]

Acetylene complexes of palladium(II) are not well understood since acetylenes are very rapidly polymerized in the presence of palladium(II) compounds.[141]

D. Palladium Carbonyls

In 1962, no simple palladium carbonyls were known.[232] In 1973, new tetracarbonyls of palladium and platinum were prepared by cocondensation of palladium and platinum with carbon monoxide at 4.2-10°K.[255] For example, cocondensation of palladium with a 1:1000 carbon monoxide-argon mixture at 10°K gave $Pd(CO)_n$ where *n*=1-3, as well as small amounts of $Pd(CO)_4$.[256]

E. Alloys

Binary alloys of palladium include those with aluminum, arsenic, boron, bismuth, chromium, copper, gallium, mercury, indium, magnesium, niobium, phosphorus, lead, sulfur, selenium, tin, strontinum, tantalum, tellurium, thorium, uranium, vanadium, zinc, and zirconium. Binary palladium alloys are used for brazing alloys, thermal fuses, and electrical contacts and resistors.[76]

Ternary alloys, which are used for dentistry and jewelry as well, include those with Ag-Au, Ag-Cu, Ag-Mn, Au-Cu, Au-Ni, Au-Pt, Cr-Ni, Cu-Ni, and Mn-Ni.[76]

Palladium absorbs 800-900 times its own volume of hydrogen. Two solution phases are probably formed rather than the pure substance Pd_2H.[17] Pd_2H is termed a metal-hydrogen alloy. Hydrogen-palladium alloys are unique, especially in the extremely rapid diffusion of hydrogen in them.[257]

Superlattice alloys of palladium, which have ordered plus disordered structure, include Cu_4Pd, FePd, $FePd_3$, BePd, and CuPd.[206] Superlattice formation occurs in systems of platinum or palladium with iron, cobalt, nickel, or copper but not in binary systems of the platinum metals themselves.[17]

APPENDIX D

ANALYSIS

The analytical chemistry of palladium is treated in detail in the 1966 publication by F. E. Beamish, The Analytical Chemistry of the Noble Metals.[243/]

Inorganic and organic palladium compexes are important in many of the separation and determination procedures. Examples are palladium dimethylglyoximate in precipitation methods; 1-nitroso-2-naphtholpalladium in solvent extraction; and hydroxo, nitrito, or chloro complexes in ion-exchange and adsorption chromatographic methods, which are most useful for analyzing amounts present in ore deposits.

Discussed briefly are methods for determining palladium by gravimetric, volumetric, fire assay, spectrochemical and spectrographic, polarographic, and neutron-activation-analysis methods.

A. Gravimetric and Volumetric Analyses

Fifty gravimetric reagents are known for palladium, the most important palladium precipitant being the oxime family. The most generally useful reagents for the precipitation of metallic palladium are hydrazine, acetylene, and ethylene.

Many of the volumetric methods for determining palladium involve formation of insoluble palladium compounds and incorporate some physical process for determining the stoichiometric end-point or involve adding excess standard complexing reagent and determining the excess by back titration. A few of the procedures are useful for routine analysis of simple alloys.

B. Fire Assay

The classical fire assay of ores involves preparing suitable fusion mixtures and then heating to give a slag, from which the noble metals are extracted by an alloying or collecting metal, which is most often lead. Flux constituents contain a small amount of a reducing agent such as flour, which produces the desired amount of lead from litharge. The slags are silicates and borates of copper, iron, and nickel oxides mutually dissolved with lead silicates and borates. The lead button, containing sufficient silver to collect the noble metals, in the cupel (vessel) is oxidized to lead oxide liquid, which wets the cupel and is absorbed by it along with the base metals such as copper and nickel. Palladium losses to the cupel are insignificant. After cupellation, the noble metals remain in the silver bead or prill.

If the amount of gold in the silver assay bead exceeds that of palladium, most, if not all, of the palladium can be dissolved with the silver in nitric acid. Gold seems to assist the dissolution of palladium and platinum in nitric acid. Palladium is determined as the dimethylglyoximate. Treating the silver bead with hot concentrated sulfuric acid dissolves most of the silver, some palladium, and none of the other metals. Palladium is removed as the hydrated oxide or its dimethylglyoximate. The contents of silver assay beads can also be solvent extracted from the acid parting solution.[243]

Base-metal-sulfide mattes may form a third phase in the final slag during the fire assay. These mattes are excellent carriers of platinum metals. Use of the matte as the sole collector is possible, but the matte must be avoided with lead as the collector. The nitre assay involves partially oxidizing the sulfides by adding potassium nitrate and using the excess of base-metal sulfides to produce the lead button.

A new assay method by Beamish makes use of the naturally occurring base metal associates iron, copper, and nickel. The method had not yet been applied directly to sulfide ores and concentrates because the constituents of the charge would produce a matte. The method employs a flux comprising carbon, sodium carbonate, and borax and uses graphite as the preferred reducing agent. The new assay reduces slag losses of platinum and palladium. Nickel in the slag from the classical assay increases the risk of some loss of platinum metals.[243]

C. Spectrochemical and Spectrographic Methods

The small amounts of noble metals in primary deposits require some type of concentration prior to any spectrophotometric determination. The lead button or silver bead can be treated readily for such determinations. Colorimetric reagents for palladium are numerous and varied. The approximately fifty methods are useful for palladium concentrations 0.05-250 ppm. These reagents include organic complexing compounds and tin(II) salts.[243] For example, p-nitrosodiphenylamine reacts with palladium salts to give a red complex, which absorbs at 510 mμ. Also used are p-nitrosodimethylaniline, EDTA, 2-nitroso-1-naphthol, α-furildioxime and p-bromoaniline.[258]

Spectrochemical analysis can be used for determining low or trace concentrations of platinum metals, trace impurities in refined platinum metals, and intermediate or high concentrations of platinum metals in alloys or mixtures.[243]

Optical emission spectrography, in which optical spectra are excited by electrical discharges, is used with fire assaying in determining low or trace concentrations (<1%) in rocks, ores, and metallurgical materials and for determining trace impurities.

X-ray emission spectrometry is more useful for higher concentrations although newer instruments allow trace determinations in favorable cases.[243]

In spark-emission spectrography, F. A. Pohl (1953-54) separated palladium in rocks, soils, ores and (or) waters by extraction as organic complexes and then evaporated the extract on a graphite electrode. Pohl determined palladium in plants after separating the dithiocarbamate, oxime, or dithizone complex by chloroform extraction.[258]

D. Polarography

Palladium gives polarographic waves, which are sensitive to a few microns per milliliter, in potassium cyanide, ammonia, ammonium chloride, pyridine, or ethylenediamine. Separation by dimethylglyoxime is necessary in the presence of platinum, rhodium, or iridium.

E. Neutron Activation Analysis

The most sensitive analytical method for palladium is radioactivation.[258] Neutron activation analysis is the method most free of reagent and laboratory contamination.[259] The detection limit for ^{109m}Pd or ^{109g}Pd (half-lives 4.8 min and 13.6 hour, respectively, γ-ray energies 0.19 Mev and 0.088 Mev) is 0.002 μg (based on 1-hour irradiation at 4.5 X 10^{13} neutrons/cm^2 sec). It can be used for analyzing blood, tissue, water, and airborne particulates.[175]

BIBLIOGRAPHY

1. Wright, T. L., and M. Fleischer, Geochemistry of the Platinum Metals, Bulletin 1214-A, Department of the Interior, U. S. Government Printing Office, Washington, D. C., 1965.

2. Mertie, J. B., Jr., Economic Geology of the Platinum Metals, Geological Survey Professional Paper 630, Department of the Interior, U. S. Government Printing Office, Washington, D. C., 1969.

3. Harriss, R. C., J. H. Crocket, and M. Stainton, "Palladium, Iridium and Gold in Deep-Sea Manganese Nodules," Geochim. Cosmochim. Acta, **32**, 1049-56 (1968).

4. Arthur D. Little, Inc., Water Quality Data Book, Vol. 2, Inorganic Chemical Pollution of Freshwater, Water Pollution Control Research Series, Environmental Protection Agency, U. S. Government Printing Office, Washington, D. C., 1971.

5. Kessler, T., A. G. Sharkey, and R. A. Friedel, Spark-Source Mass Spectrometer Investigation of Coal Particles and Coal Ash, Bureau of Mines Progress Report No. 42, U. S. Department of the Interior, Washington, D. C., 1971.

6. Swaine, D. J., The Trace-Element Content of Fertilizers, Commonwealth Agr. Bureau, Farnham Royal, Bucks, England, 1962.

7. Charles River Associates, Inc., Economic Analysis of the Platinum Group Metals, Prepared for Property Management and Disposal Services, General Services Administration, Washington, D. C., 1968.

8. Ageton, R. W., and J. P. Ryan, "Platinum-Group Metals" in Minerals Facts and Problems, Bureau of Mines Bulletin 650, U. S. Department of the Interior, U. S. Government Printing Office, Washington, D. C., 1970, pp. 653-669.

9. Ware, Glen C., "Platinum Group Metals" in Mineral Facts and Problems, Bureau of Mines Bulletin 630, U. S. Department of the Interior, U. S. Government Printing Office, Washington D. C., 1965, pp. 711-719.

10. Petrick, A., Jr., H. J. Bennett, K. E. Starch, and R. C. Weisner, The Economics of Byproduct Metals (in Two Parts). 1. Copper System. Information Circular 8569, U. S. Department of the Interior, Bureau of Mines, U. S. Government Printing Office Stock No. 2404-01341, Washington, D. C., 1973.

11. Gonser, B. W., Battelle Memorial Institute, Columbus, Ohio, personal communication, December 1973.

12. Martin, H., Martin Metals, Member of Precious Metals Commission of National Association of Secondary Metal Industries, personal communication, December 1973.

13. Anonymous, "Lead Refining," SARCO News, 12 (1), 7-11 (1972).

14. Rosenson, R., "Importance of Precious Metals Recovery," Secondary Raw Materials, 9 (5), 77-9 (1971).

15. Luning, D., Vice President of Marketing, Matthey Bishop, Inc., personal communication, December 1973.

16. Anonymous, "Market Newsletter, 'Palladium Prices were Hiked $2/Troy Ounce Last Week'" Chem. Week., 102 (7), 49-50 (1968).

17. Beamish, F. E., W. A. E. McBryde, and R. R. Barefoot, "The Platinum Metals" in Rare Metals Handbook, 2nd ed., C. A. Hampel, Ed., Reinhold Publishing Corporation, Chapman and Hall, Ltd., London, 1961, pp. 304-334.

18. Mitko, Francis C., "Platinum-Group Metals" in Minerals Yearbook 1970, Vol. I, Metals, Minerals, and Fuels, Bureau of Mines, U. S. Department of the Interior, U. S. Government Printing Office Stock No. 2404-1126, Washington, D. C., 1972, pp. 939-950.

19. Anonymous, "Precious Metals Become Dearer," Chem. Week, 102 (20), 49, 51 (1968).

20. Bard, J., Matthey Bishop, Inc., Malvern, Pennsylvania , personal communication, April 1973.

21. Johnston, C., "Platinum Mining in Alaska. Dredge and Dragline Operations at Goodnews Bay," Platinum Metals Rev., 6 (1), 68-74 (1962).

22. Illis, A., B. J. Brandt, and A. Manson, "The Recovery of Osmium from Nickel Refinery Anode Slimes," Metallurgical Trans., 1 (2), 431-4 (1970).

23. Anonymous, The Mond Platinum Metals Refinery, The Mond Nickel Company, Ltd., London, England [undated].

24. Gouldsmith, A.F.S., and B. Wilson, "Extraction and Refining of the Platinum Metals. A Complex Cycle of Smelting, Elecrolytic and Chemical Operations," Platinum Metals Rev., **7** (4), 136-143 (1963).

25. White, L., "The Newer Technology: Where It Is Used and Why," *Chem. Eng.*, **80** (9), AA-CC (1973).

26. Schack, C. H., and B. H. Clemmons, *Review and Evolution of Silver-Production Techniques*, U. S. Bureau of Mines Information Circular 8266, U. S. Goverment Printing Office, Washington, D. C., 1965.

27. Arthur G. McKee & Company, "Va. Copper, Zinc, and Lead Smelting Practice," *Systems Study for Control of Emissions. Primary Nonferrous Smelting Industry*, Vol. I, National Air Pollution Control Administration, Public Health Service, U. S. Department of Health, Education, and Welfare, 1969, pp. Va-1, Vb-5.

28. Reno, H. T., "Nickel" in *Minerals Facts and Problems*, Bureau of Mines Bulletin 650, U. S. Department of the Interior, U. S. Government Printing Office, Washington, D. C., 1970, pp. 347-360.

29. Mitko, F. C., "Platinum-Group Metals" in Minerals Yearbook 1971, Vol. I, *Metals, Minerals, and Fuels*, Bureau of Mines, U. S. Department of the Interior, U. S. Government Printing Office, Washington, D. C., 1973.

30. Jeffries, H. D., "Inorganic Pigments" in *Dispersion of Powders in Liquids*, G. D. Parfitt, Ed., Elsevier Publishing Co., New York, N. Y., 1969, pp. 285-324.

31. Schmidt-Fellner, A., "Role of Refiner in Precious Metals," *Secondary Raw Materials*, **9** (5), 64-70 (1971).

32. Johnson, O. C., "Chapter 5. Refining Processes" in *Silver. Economics, Metallurgy, and Use*, A. Butts and C. D. Coxe, Eds., D. Van Nostrand Co., Inc., Princeton, N. J., 1967, pp. 57-77.

33. Carsillo, N. F., Manager Purchases and Sales Platinum-Group Metals Engelhard Industries, personal communication, December 1973.

34. Materials Advisory Board, *Technological Influence on Usage of Platinum and Palladium*, Publication MAB-247, National Research Council, National Academy of Sciences--National Academy of Engineering, Washington, D. C., 1968.

35. Anderson, R. C., and R. E. Harr (Western Electric Co., Inc.), "Recovering Precious Metal Electrical Contacts from Base Metal Supports," U. S. Patent 2,797,194 (1957); Chem. Abstr., 51, 12714a (1957).

36. Singleton, E. L., and T. A. Sullivan, "Electronic Scrap Reclamation," J. Metals, 25 (6), 31-34 (1973).

37. Anonymous, "Platinum-Group Will Clean Up in Auto Market," Chem. Week, 113 (12), 27-28 (1973).

38. Wight, C. G., and R. H. Hass (Union Oil Co. of California), "Hydrocracking Catalyst Regeneration of Crystalline Zeolite Composite," U. S. Patent 3,197,397 (1965), 4 pp.; Chem. Abstr., 63, 11219f (1965).

39. Wight, C. G., and R. H. Hass (Union Oil Co. of California), "Hydrocracking Catalyst Regeneration," U. S. Patent 3,197,399 (1965) 4 pp.; Chem. Abstr. 63, 8100g (1965).

40. Clement, W. H., and C. M. Selwitz (Gulf Research and Development Co.), "Recovery of Platinum Group Metal Catalyst," U. S. Patent 3,303,020 (1967), 6 pp.; Chem. Abstr., 67, 2018 (1967).

41. Atlantic Richfield Co., "Solution for Recovering Contaminated Metal Catalysts," French Patent 2,025,331 (1970), 12 pp.; Chem. Abstr., 74, 270 (1971).

42. Sargent, H. (Atlantic Richfield Co.), "Recovery [of] Platinum-Group Metal Salts from Catalysts Contaminated with Copper," U. S. Patent 3,488,144 (1970), 3 pp.; Chem. Abstr., 72, 290 (1970).

43. Laporte Chemicals Ltd., "Reclaiming Noble Metals from Spent Catalysts," Belgian Patent 653,225 (1965), 11 pp.; Chem. Abstr., 64, 10452h

44. Riemenschneider, W., O. E. Baender, and V. Schwenk, (Farbwerke Hoechst A.-G.), "Regeneration of Catalysts for Olefin Oxidation," German Patent 1,143,499 (1963), 2 pp.; Chem. Abstr., 59, 5839h (1963).

45. Goss, H. F., and R. A. DeJean (Eastman Kodak Co.), "Recovery of Palladium from Solution," Ger. Offen. 1,816,162 (1969); Chem. Abstr., 71, (1969).

46. Fetscher, C. A. (NOPCO Chemical Co.), "Metal Extraction," U. S. Patent 3,088,799 (1963), 13 pp., Chem. Abstr., 59, 3642e (1963).

47. Rylander, P. N., Catalytic Hydrogenation over Platinum Metals, Academic Press, New York, N. Y., 1967.

48. Lait, R., and D. R. Lloyd-Owen (Laporte Chemicals Ltd.), "Recovery of Palladium from Catalysts," British Patent 922,021 (1963), 6 pp.; Chem. Abstr., 59, 240e (1963).

49. Nixon, W. G. (Universal Oil Products), "Recovery of Platinum from Composite Catalysts Containing Platinum and Alumina," U. S. Patent 2,828,200 (1958); Chem. Abstr., 52, 11,321b (1958).

50. Kolyer, J. M. (FMC Corp.), "Reactivation of Spent Pd-C Hydrogenation Catalyst," U. S. Patent 3,214,385 (1965); 3 pp.; Chem. Abstr., 64, 1969g (1966).

51. Goss, H. F. (Eastman Kodak Co.,), "Recovery of Palladium Deposited on Surfaces," Ger. Offen. 1,926,042 (1969), 11 pp.; Chem. Abstr., 72, 350 (1970).

52. Hirschberg, R. (Farbwerke Hoechst (A.-G.), "Recovery of Palladium from Acidic Solutions, " German Patent 1,205,069 (1965), 3 pp.; Chem. Abstr., 64, 4652c (1966).

53. Carlson, G. A. (PPG Industries, Inc.), "Recovery of Metals From Trace Amounts of Metal Ions in Solution by Electrolysis," Ger. Offen. 2,011,610 (1970), 46 pp.; Chem. Abstr., 74, 200 (1970).

54. Philpott, J. E., "Palladium Plating of Printed Circuits," Platinum Metals Rev., 4, 12-14 (1960).

55. Lowe, W., "Origin and Characteristics of Toxic Wastes with Particular Reference to the Metal Industries," J. Water Pollution Control Federation, 69 (3), 270-280 (1970).

56. Panesko, J. V., "Recovery of Rhodium, Palladium, and Technetium on Strongly Basic Anion Exchange Resin," U. S. At. Energy Comm., ARH-1279 (1969).

57. Panesko, J. V., "Process for Recovery of Rhodium, Palladium, and Technetium from Aged Reprocessing Wastes at Hanford," U.S. At. Energy Comm., ARH-733 (1968).

58. Rohrmann, C. A., "Large Scale Exploitation of Fission Product Rhodium and Palladium," Engelhard Ind. Tech. Bull., 9 (2), 56-61 (1968).

59. Hoyt, C. D., and J. P. Ryan, "Platinum-Group Metals" in Minerals Yearbook, 1969, Vol. I-II, Metals, Minerals, and Fuels, Bureau of Mines, U. S. Department of the Interior, U. S. Government Printing Office, Washington, D. C., 1970.

60. Colvin, C. A., "Recovery of Palladium from Nuclear Waste Solutions Using a Packed Column of Tricapryl Monomethyl Ammonium Chloride on an Inert Support," U. S. At. Energy Comm., ARH-SA-28 (1969).

61. Pepkowitz, L. P., B. L. Vondra, W. C. Judd, J. E. Scott, G. G. Erlich, and E. R. Shuster, "Decontamination and Recovery of Precious Metals," U. S. At. Energy Comm. NYO-9175 (1962); Chem. Abstr., **56**, 1391d (1962).

62. Anonymous (F.M.L.), "Chemistry of the Platinum Metals. A Review of the American Chemical Society Meeting," Platinum Metals Rev., **14** (1), 21-23 (1970).

63. Newman, R. J., and F. J. Smith, "Platinum Metals from Nuclear Fission. An Evaluation of Their Possible Use by Industry," Platinum Metals Rev., **14** (3), 88-92 (1971).

64. Lyman, T., H. E. Boyer, E. A. Durand, W. J. Carnes, M. W. Chevalier, H. C. Doepken, P. D. Harvey, H. Lawton, T. M. Leach, I. A. Anderson, H. V. Bukovics, B. A. Caldwell, and E. R. Watkins, Eds., Metals Handbook, Vol 4, Forming, 8th ed., American Society for Metals, Metals Park, Ohio, 1969.

65. Carter, F. E., "The Noble Metals Find Increasingly Wide Use in Industry," Mater. Methods, **28** (5), 55-59 (1948).

66. Jahn, C. A. H., "Platinum Metals. A Survey of Their Production, Properties and Engineering Uses. Part IV--Alloys," Metal Ind. (London), **72**, 267-269 (1948).

67. Philpott, J. E., "Applications of the Noble Metals in the Chemical Industry," S. African Chem. Processing, **4** (2), CP12-CP15 (1969).

68. Safranek, W. H., "Introductory Data on Properties and Applications' in Symposium on Electrodeposited Metals as Materials for Selected Applications, Nov. 3-4, 1971, Metals and Ceramics Information Center, Battelle Columbus Labs., Columbus, Ohio, 1972, pp. 1-10.

69. Lyman, T., H. E. Boyer, P. M. Unterweise, J. E. Foster, J. P. Hontas, and H. Lawton, Eds., Metals Handbook, Vol. I., Properties and Selections of Metals, 8th ed., American Society for Metals, Novelty, Ohio, 1961.

70. Vivian, P. G. L., "Precious Metal Plaing," Electroplating Metal Finishing, 23 (11), 20-28 (1970).

71. Rhys, D. W., and W. Betteridge, "Brazing for Elevated Temperature Service," Metal Ind., 101 (2), 2-4, 27-30, 45-46 (1962).

72. Standen, A., Executive Ed., "Platinum-Group Metals" in Kirk-Othmer Encyclopedia of Chemical Technology, Vol. 15., 2nd ed., Interscience Publisher, Division John Wiley & Sons, New York, N. Y., 1963, pp. 852-870.

73. Wise, E. M., Palladium Recovery, Properties and Uses, Academic Press, New York, N. Y., 1968.

74. Cramer, S. D., and D. Schlain, "Electrodeposition of Palladium and Platinum from Aqueous Electrolytes," Plating, 56, 516-522 (1969).

75. Vivian, P. G. L., "Advances in Precious Metal Plating," Prod. Finishing (London), 17 (11), 97-101 (1964).

76. Anonymous, Palladium, The Metal, Its Properties and Application, The International Nickel Company, Inc., New York, N. Y., 1965 [ ].

77. Anonymous, "Electroplating the Platinum Metals," Platinum Metals Rev., 14 (3), 93-94 (1970).

78. Hunt, L. B., and F. M. Lever, "Availability of the Platinum Metals," Platinum Metals Rev., 15 (4) 126-138 (1971).

79. Krohn, A., and C. W. Bohn, "Electrodeposition of Alloys: Present State of the Art," Electrodeposition and Surface Treatment, 1 (3), 199-211 (1973).

80. Pearlstein, F., and R. F. Weightman, "Electroless Palladium Deposition," Plating, 56, 1158-1161 (1969).

81. Philipp, W. H., and R. A. Lad, "Radiation-induced Preparation of Pure Metals from Solution," NASA Spec. Publ. 1969 (Publ. 1970), NASA SP-227, 229-237; Chem. Abstr., 73, 189 (1970).

82. Anonymous, The Platinum-Group Metals in Industry, The International Nickel Company, Inc., New York, N. Y., 1963 [ ].

83. Darling, A. S., "Platinum Coatings on Base Metals and Refractory Materials," Chem. Ind., 1973 (19), 928-930 (1973).

84. Vines, R. F., "Chapter 22. Noble Metals" in Corrosion Resistance of Metals and Alloys, 2nd ed., F. L. LaQue and H. R. Copson, Eds., American Chemical Society Monograph No. 158, Reinhold Publishing Corp., New York, N. Y., 1963, pp. 601-622.

85. Peckner, D., "Precious Metals and Their Uses," Mater. Design Eng., 57 (6), 93-102 (1963).

86. Perygi, P., American Metal Climax, Inc., personal communication, 1973.

87. Burke, D. P., "Catalysts. Part 1. Petroleum Catalysts" and "Part 2. Chemical Catalysts," Chem. Week, 111 (18), 23-33 (1972) and 111 (19), 35-45 (1972).

88. Augustine, R. L., Catalytic Hydrogenation, Marcel Dekker, Inc., New York, N. Y., 1965.

89. Beamer, R. L., R. H. Belding, and C. S. Fickling, "Stereospecific Hydrogenations III: Palladium-on-Poly-S-Valine and Palladium-on-Poly-S-Leucine," J. Pharm. Sci., 58 (11), 1419-1421 (1969).

90. Beamer, R. L., R. H. Belding, and C. S. Fickling, "Stereospecific Hydrogenations IV: Palladium-on-Poly-S-Valine and Palladium-on-Poly-S-Leucine," J. Pharm. Sci., 58 (11), 1419-1421 (1969).

91. Martin, E. W., E. F. Cook, E. E. Leuallen, A. Osol, L. F. Tice, and C. T. Van Meter, Eds., Remington's Practice of Pharmacy, 12th ed., Mack Publishing Co., Easton, Pa., 1961.

92. Anonymous, "The Story of the Platinum Metals," Engelhard Industries, Inc., [undated].

93. Dutton, W. L., and R. C. Hirt, "Ultraviolet Emission Spectrographic Determination of Trace Palladium to Ascertain the Method Used to Produce Tetracycline," J. Pharm. Sci., 51, 911 (1962).

94. Schulte, K. E., G. Henke, and K. S. Tjan, "Determination of Metal Traces in Antibiotics by Neutron Activation Analysis," Pharm. Acta Helv., 45, 265-281 (1970).

95. Martin, E. W., Ed.-in-Chief, Remington's Pharmaceutical Sciences, 13th Ed., Mack Publishing Co., Easton, Pa., 1965.

96. Stecher, P. G., Ed., The Merck Index, 8th ed., Merck and Co., Inc., Rahway, N. J., 1968.

97. Warner, P. O., and H. F. Barry, Molybdenum Catalyst Bibliography (1950-1964), Climax Molybdenum Co., an AMAX Subsidiary, New York, N. Y., 1967.

98. Anonymous, "1972 Refining Processes Handbook," Hydrocarbon Process., **51** (9), 111-222 (1972).

99. Standen, Anthony, Executive Ed., "Petroleum Refinery Processes," Kirk-Othmer Encyclopedia of Chemical Technology, 2nd ed., Vol. 15, Interscience Publishers, a Division of John Wiley and Sons, New York, N. Y., 1968, p. 55.

100. Acres, G. J. K., B. S. Cooper, and G. L. Matlack, "The Production of Automobile Emission Control Catalysts. Global Capacity of the Johnson Matthey Group," Platinum Metals Rev., **17** (3), 82-87 (1973).

101. Iammartino, N. R., "Detroit's Catalyst Choices," Chem. Eng., **80** (27), 24-26 (1973).

102. Anonymous, "Extension of Emission Standards Affects Auto Giants Differently, Ford Puts Off Carb, Cancels Tool Orders," American Metal Market/Metal Working News, **80** (248), 1, 10 (1973).

103. McIlheny, V. K., "Danger Feared in Catalysts to Control Car Pollution," The Kansas City Times, p. 6B (November 23, 1973).

104. Colburn, D. A., "A Cure Worse Than the Disease?," Automotive Industries **149** (8), 15 (1973).

105. Anonymous, "The Idea of Putting Catalytic Emission-Coverters on U. S. Cars is Getting Buffeted," Chem. Eng., **80** (25), 59-20 (1973).

106. Colburn, D. A., "Pollutant Uncertainties at EPA," Automotive Industries, **149** (11), 20 (1973).

107. Anonymous, "Meanwhile, EPA is Starting to Fret over Another Possible Health Threat from Autos," Chem. Eng., **80** (16), 30 (1973).

108. Gottesman, C. A., "NAS Report on Current Emission Control Technology Sent to Congress and the EPA," Automotive News, 148 (6), 17-18 (1973).

109. Callahan, J. M., "Catalytic Converters: Will the Seed Money Pay Off?," Inter. Automotive Ind., 148 (8), 33-38 (1973).

110. Jagel, K. I., "HC/CO Oxidation Catalysts for Vehicle Exhaust Emission Control," American Petroleum Institute Proceedings, Division of Refining, 169-185 (1971).

111. Panel on Catalysts for Automotive Emission Devices and Petroleum Refining of the Committee on Technical Aspects of Critical and Strategic Materials, National Materials Advisory Board, Substitute Catalysts for Platinum in Automobile Emission Control Devices and Petroleum Refining, Publication NMAB-297, National Academy of Sciences--National Academy of Engineering, Washington, D. C., 1973.

112. Cole, E. N., "How Detroit and Washington View Vehicle Regulations. The Anatomy of a Decision," Automotive Eng., 81 (12), 52, 54, 56, 58, 60 (1973).

113. Callahan, J. M., "First Details of GM's '75 Emission Control System," Automotive News, 148 (5), 36-38 (1973).

114. Anonymous, "Stalling Time Has Run Out for Chrysler," Am. Metal Market/ Metal Working News, 80 (233), 5 (1973).

115. Callahan, J. M., "GM Catalyst Systems on Display," Inter. Automotive Ind., 149 (6), 18-19 (1973).

116. Anonymous, "Latest Auto Exhaust Catalyst Supplier," Chem. Eng. News, 51 (32), 13 (1973).

117. Anonymous, "Auto Makers Set Up for Catalytic Systems," Chem. Eng. News, 50 (50), 2-3 (1972).

118. Johnson, D. E. L., "Large Platinum, Palladium Imports Seen Cut by Emission Device Salvage," Am. Metal Market/Metal Working News, 80 (181), 2-3 (1972).

119. Fosdick, R. J., "Converter Contract Line-Up Grows Longer As Suppliers Jockey for '75 Positions," Inter. Automotive Ind., 148 (2), 15 (1973).

120. Suskind, D. A., "Platinum-Group, A Bonanza in the Automotive Industry," Eng. Mining J., 174, 83-85 (1973).

121. Fosdick, R. J., "Catalysts Looking Better to GM," Intern. Automotive Ind., 149 (2), 17 (1973).

122. Belous, R., "Converter Ire Still Burns," Am. Metal Market/Metal Working News, 80 (219), 42 (1973).

123. Shelef, M., and H. S. Gandhi, "Ammonia Formation in Catalytic Reduction of Nitric Oxide by Molecular Hydrogen," Ind. Eng. Chem. Prod. Res. Develop., 11 (4), 393-396 (1972).

124. Balgord, W. D., "Fine Particles Produced from Automotive Emissions Control Catalysts," Science, 180, 1168-1169 (1973).

125. Ross, R. D., Ed., Industrial Waste Disposal, Reinhold Book Corporation, New York, N. Y., 1968.

126. Cohn, J. G., "Precious Metals Withstand Tough Environments," Mater. Design Eng., 64 (3), 100-105 (1966).

127. McCormick, J. C. (FMC Corp.), "Hydrogen Peroxide Decomposition Catalyst Coated with Samarium Oxide," U. S. Patent 3,560,407 (1971), 3 pp.; Chem. Abstr., 74, 276 (1971).

128. Mikovsky, R. J., and S. Srinivasan (Mobil Oil Corp.) "Production of Heavy Water Employing a Group VIII Catalyst," U. S. Patent 3,681,021 (1972), 5 pp.; Chem. Abstr., 77, 488 (1972).

129. Cooper, G. B., "Gas Packing," Food Manuf., 39 (2), 49-50 (1964); Chem. Abstr., 60, 13787a (1964).

130. Metal Box Co., Ltd., "Method to Remove the Remaining Oxygen in Containers," Netherlands Patent 64-00122 (1964), 5 pp.; Chem. Abstr. 62, 1017b (1965).

131. Quesda, C., and R. W. Neuzil (Universal Oil Products), "Oxygen Scavenging from Closed Containers," U. S. Patent 3,437,428 (1969), 4 pp.; Chem. Abstr., 70, 283 (1969).

132. Darling, A. S., "The Vapour Pressures of the Platinum Metals. A Review of Some Recent Determinations," Platinum Metals Rev., 8 (4), 134-140 (1964).

133a. W. E. Davis and Associates, National Inventory of Sources and Emissions. Barium, Boron, Copper, Selenium, and Zinc. Copper, Section III., Environmental Protection Agency, Research Triangle Park, N. C., 1972.

133b. Schack, C. H., and B. H. Clemmons, "Chapter 4, Extractive Processes" in Silver. Economics, Metallurgy, and Use, A. Butts and C. D. Coxe, Eds., D. Van Nostrand Co., Inc., Princeton, N. J., 1967, pp. 57-77.

134. Doudoroff, P., and M. Katz, "Critical Review of Literature on the Toxicity of Industrial Wastes and their Components to Fish. II. The Metals, as Salts," Sewage Ind. Wastes, **25**, 802-839 (1953).

135. Wood, J., "3-Year Platinum Fugure Promising; After That, It's Anyone's Guess," Am. Metal Market/Metal Working News, **80** (209), 21 (1973).

136. Anonymous, Price Quotations, Am. Metal Market/Metal Working News, **80** (252), 31 (1973).

137. Anonymous, "American Metal Market Closing Prices. Friday, February 15, 1974," Am. Metal Market/Metal Working News, **81** (34), 35 (1974).

138. Faith, W. L., D. B. Keyes, and R. L. Clark, Industrial Chemicals, 3rd ed., John Wiley & Sons, Inc., New York, N. Y., 1965.

139. Anonymous, 1972 Directory of Chemical Producers United States of America, Chemical Information Services, Stanford Research Institute, Menlo Park, California, 1972.

140. Hatch, "homogeneous Catalysis for Liquid Phase Oxidation," Hydrocarbon Process., **49** (3), 101-105 (1970).

141. Hartley, F. R., "Olefin and Acetylene Complexes of Platinum and Palladium," Chem. Rev., **69** (6), 799-844 (1969).

142. Moran, J., Environmental Protection Agency, Research Triangle Park, personal communication, July 1973.

143. Anonymous, 1972 Automobile Facts and Figures, Motor Vehicle Manufacturers Association of the U. S., Inc., Detroit, Mich., 1973.

144. Keefer, H. J., and R. H. Gumley, "Relay Contact Behavior under Non-Eroding Circuit Conditions," Bell System Tech. J., **37**, 777-814 (1958).

145. Anonymous, "Training in Use Called Main Problem for Dental Alloys," Am. Metal Market/Metal Working News, **80** (73), 27 (1973).

146. Browning, Ethel, Toxicity of Industrial Metals, Butterworths, London, 1961.

147. Meek, S. F., G. C. Harrold, and C. P. McCord, "The Physiologic Properties of Palladium and Its Compounds," Ind. Med., **12**, 447-448 (1943).

148. Munro-Ashman, D., D. D. Munro, and T. H. Hughes, "Contact Dermatitis from Palladium," Trans. St. Johns Hospital Dermitol. Soc., **55**, 196-197 (1969).

149. Durbin, P. W., K. G. Scott, and J. G. Hamilton, "The Distribution of Radioisotopes of Some Heavy Metals in the Rat," Univ. Calif. (Berkeley) Publs. Pharmacol., **3** (1), 1-34 (1957).

150. Dodds, E. C., R. L. Noble, H. Rinderknecht, and P. C. Williams, "Prolongation of Action of the Pituitary Antidiuretic Substances and of Histamine by Metallic Salts," Lancet, **2**, 309-311 (1937).

151. Schroeder, H. A., and M. Mitchener, "Scandium, Chromium(VI), Gallium, Yttrium, Rhodium, Palladium, Indium in Mice: Effects on Growth and Life Span," J. Nutr., **101**, 1431-1437 (1971).

152. Schroeder, H. A., Metallic Micronutrients and Intermediary Metabolism, Final Progress Report, U. S. Army Medical Research and Development Command, Document No. AD 708581, National Technical Information Service, U. S. Department of Commerce, Springfield, Va., 1970.

153. Strokinger, H. E.,"Chapter XXVII--The Metals (excluding Lead)" in Industrial Hygiene and Toxicology, Vol. II, 2nd revised ed., Frank A. Patty, Ed., Interscience Pub., New York, N. Y., 1958.

154. Fujita, Sotoo, "Silver-Palladium-Gold Alloys Carcinogencity and Acid Mucopolysaccharides in the Induced Tumors," Shika Igaky, **34** (6), 918-932 (1971); Chem. Abstr., **77**, 110 (1972).

155. Habu, Tetsuya, "Histopathological Effects of Silver-Palladium-Gold Alloy Implantation on the Oral Submucous Membranes and Other Organs," Shika Igaku, **31** (1), 17-48 (1968); Chem. Abstr., **69**, 8885 (1968).

156. Ariel, I. M., and J. S. Robertson, "The Dose Rate Factor from Interstitial Irradiation with ^{103}Palladium and ^{109}Palladium," Surg., Gynecol. Obstet., **132**, 817-831 (1971).

157. Fawwaz, R. A., W. Hemphill, and H. S. Winchell, "Potential Use of ^{109}Pd-Porphyrin Complexes for Selective Lymphatic Ablation," J. Nucl. Med., **12** (5), 231-236 (1971).

158. Lewis, R. W., R. W. Mack, M. J. Dennis, and J. J. Woods, "Leukopenia-Induction Capacity of 6-MP Palladium Complex in the Chick," J. Proc. Soc. Exptl. Biol. Med., **131**, 1219-1222 (1969).

159. Kirschner, S., Y.-K. Wei, D. Francis, and J. G. Bergman, "Anticancer and Potential Antiviral Activity of Complex Inorganic Compounds," J. Med. Chem., **9** (3), 369-372 (1966).

160. Somers, E., "Plant Pathology: Fungitoxicity of Metal Ions," Nature, **184** (4684), 475-476 (1959).

161. White, J., "Inhibitory Effect of Common Elements towards Yeast Growth," J. Inst. Brewing, **57** (**48**, new series), 175-179 (1951).

162. Tachibana, S., T. Murakami, and S. Sawada, "Studies on CO_2-fixing Fermentation. XXIII. Effects of Trace Elements on L-Malate Fermentation," Hakko Kagaku Zasshi (J. Ferment. Technol.), **50** (3), 171-177 (1972).

163. Sarwar, M., R. J. Thibert, and W. G. Benedict, "Effect of Palladium Chloride on the Growth of *Poa pratensis*," Can. J. Plant, Sci., **50** (1), 91-96 (1970).

164. Benedict, W. G., "Some Morphological and Physiological Effects of Palladium on Kentucky Bluegrass," Can. J. Bot., **48** (1), 91-93 (1970).

165. Spikes, J. D., and C. F. Hodgson, "Enzyme Inhibition by Palladium Chloride," Biochem. Biophys. Res. Commun., **35** (3), 420-422 (1969).

166. Christensen, G. M., "Effects of Metal Cations and Other Chemicals upon the *In Vitro* Activity of Two Enzymes in the Blood Plasma of the White Sucker," Chem.-Biol Interactions, **4** (5), 351-361 (1972).

167. Shishniashvili, D. M., V. N. Lystsov, and Yu.Sh. Moshkovskii, "Vliyanie pH i ionnoi sily na degradatsiyu DNK pod deistviem palladiya," Soobsch. Akad. Nauk Gruz. SSR, 65 (2), 457-460 (1972); Chem. Abstr., 77, 149 (1972).

168. Ridge, J. D., Ed., Ore Deposits of the United States, 1933-1967, The Graton-Sales Volume, 1st ed., Vol. II, The American Institute of Mining, Metallurgical, and Petroleum Engineers, Inc., New York, N. Y., 1968.

169. Parker, Raymond L., "Chapter D. Composition of the Earth's Crust" in Data of Geochemistry, 6th ed., Michael Fleischer, Tech. Ed., Geol. Survey Prof. Paper 440-D, U. S. Government Printing Office, Washington, D. C., 1967.

170. Vinogradov, A. P., "Average Content of Chemical Elements in the Main Types of Igneous Rocks in the Earth's Crust," Geokhimiya, 1962b (7), 555-571 (1962); Geochemistry, 1962 (7), 641-664 (1963).

171. Mitchell, R. L., "Chapter 9. Trace Elements" in Chemistry of the Soil, F. E. Bear, Ed., Reinhold Publishing Corp., New York, N. Y., 1955, pp. 253-285.

172. MacDougall, J. D., and R. C. Harriss, "The Geochemistry of an Arctic Watershed," Can. J. Earth Sci., 6 (2), 305-315 (1969).

173. Page, N. J., L. B. Riley, and J. Haffty, Platinum, Palladium, and Rhodium Analyses of Ultramafic and Mafic Rocks from the Stillwater Complex, Montana, Geological Survey Circular No. 624, U. S. Geological Survey, U. S. Department of the Interior, Washington, D. C., 1969.

174. Foster, R. L., "Nickeliferous Serpentinite near Beaver Creek, East-central Alaska" in Some Shorter Mineral Resource Investigations in Alaska, Geological Survey Circular 615, U. S. Geological Survey, U. S. Department of the Interior, Washington, D. C., 1969, pp. 2-4.

175. Hawley, J. E., and Y. Rimsaite, "Platinum Metals in Some Canadian Uranium and Sulfide Ores," Am. Mineralogist Washington, 38, 463-475 (1953).

176. Keays, R. R., and H. Crocket, "A Study of Precious Metals in the Sudbury Nickel Irruptive Ores," Econ. Geol., 65 (4), 438-450 (1970).

177. Beamish, F. E., and M. G. Bapat, "Simultaneous Determination of Microgram Amounts of Platinum and Palladium in the Presence of Base Metals," Anal. Letters, 2 (7), 387-394 (1969).

178. Tomingas, N., and W. C. Cooper, "Spectrographic Determination of Palladium in Dore' Metal," Appl. Spectr., 11 (4), 164-166 (1957).

179. Year Book of the American Bureau of Metal Statistics, 51st Annual Issue for the Year 1971, American Bureau of Metal Statistics, New York, N. Y., 1972.

180. Anonymous, "Imports Commodity by Country "[1971 and 1972] and "U. S. Imports. General and Consumption, Schedule A. Commodity and Country" [1968-1970], Bureau of the Census, U. S. Department of Commerce Foreign Trade Series FT 135 (1968-1972).

181. American Metal Market, Metal Statistics 1972, Fairchild Publications, Inc., New York, N. Y., 1972.

182. Heindl, R. A., "Platinum-Group Metals" in Minerals Yearbook, Volume I-II, Metals, Minerals, and Fuels, Bureau of Mines, U. S. Department of the Interior, U. S. Government Printing Office, Washington, D. C., 1967.

183. Yearbook of the American Bureau of Metal Statistics, 49th Annual Issue for the Year 1969, American Bureau of Metal Statistics, New York, N. Y., 1970, p. 35.

184. Cordero, H. G., and T. J. Tarring, Eds., Non-Ferrous Metal Works of the World, 1st ed., Metal Bulletin Books, Ltd., London, 1967.

185. Anonymous, "Palladium" in Thomas Register of American Manufacturers and Thomas Register Catalog File 1971, 61st ed., Vol. 4, Products and Services, Thomas Publishing Co., New York, N. Y., 1971, pp. 5648-5649.

186. Chemical Information Services, 1973 Directory of Chemical Producers. United States of America. Stanford Research Institute, Menlo Park, Calif., 1973, p. 727.

187. Anonymous, "The New Post Office Relay (The Reliability of Palladium-Silver Contacts)," Platinum Metals Rev. 15 (4), 141 (1971).

188. Anonymous, Precious Metals, Electrical Contacts, and Contact Components, The J. M. Ney Co., Bloomfield, Connecticut, 1972 [?].

189. Vines, R. F., "Precious Metal Electrodeposits for Electrical Contact Service," Plating, 59, 923-925 (1967).

190. Manzone, M. G., and J. Z. Briggs, Mo. Less-Common Alloys of Molybdenum, Climax Molybdenum Company, New York, N. Y., 1962.

191. Luning, D., "How to Buy Platinum-Group-Metals Chemicals," Chem. Eng. 79 (28), 114-118 (1972).

192. Tugwell, G. L., "Industrial Applications for the Noble Metals," Metal Prog., 88, 73-78 (1965).

193. Ellern, H., Military and Civilian Pyrotechnics, Chemical Publishing Co., Inc., New York, N. Y., 1968.

194. Stecher, P., M. J. Finkel, O. H. Siegmund, and B. M. Szafranski, Eds., The Merck Index of Chemicals and Drugs, 7th ed., Merck and Co., Inc., Rahway, N. J., 1960.

195. Stiles, D. A., and P. H. Wells, "The Production of Ultra-pure Hydrogen," Platinum Metals Rev., 16 (4), 124-128 (1972).

196. Anonymous, "Minature Palladium Diffusion Tubes for Chromatography," Platinum Metals Rev., 15 (4), 131 (1971).

197. Uhlig, H. H., Ed., The Corrosion Handbook, John Wiley and Sons, Inc., New York, N. Y., 1948.

198. Anonymous (based on a paper by E. Wicke), "Isotope Effects in Palladium-Hydrogen Systems. The Concentration of Deuterium and Tritium," Platinum Metals Rev., 15 (4), 144-146 (1971).

199. Anonymous (L.L.S.), "Surface Treatment of Titanium with Palladium," Platinum Metals Rev., 14 (2), 47 (1970).

200. Burton, C., "Production of Focal Brain Lesions by Inductive Heating," U. S. Patent 3,653,385 (1972), 7 pp.; Chem. Abstr. 77, 95 (1972).

201. Foulke, D. G., "Plating on Precious Metals," Plating, 51 (12), 685-688 (1964).

202. Rice, J. M., and I. J. Hutkin, (Dynasciences Corp.) "High-Temperature Magnetic Recording Tape," U. S. Patent 3,607,149 (1971), 3 pp.; Chem. Abstr. 72, 363 (1972).

203. Fisher, R. D., and W. H. Chilton (National Cash Register Co.), "Magnetic Memory Device", French Patent 1,494,807 (1967), 8 pp.; Chem. Abstr. 69, 8560 (1969).

204. Driscoll, J. S., and C. N. Matthews (Monsanto Research Corp.), "Complex Triphenylphosphorane Metal Polyhalides", U. S. Patent 3,374,256 (1968), 11 pp.; Chem. Abstr. 69, 3392 (1968).

205. Neblette, C. B., Ed., Photography, Its Materials and Processes, 6th ed., D. Van Nostrand and Co., Inc., New York, N. Y., 1962.

206. Fuji Photo Film Co. Ltd., "Platinum-Group Metal-Quinone Complexes," U. S. Patent 3,656,961 (1973); Platinum Metals Rev., 17 (1), 44 (1973).

207. Rasch, A. A. (Eastman Kodak Co.), "Image Receptor Layers for the Photographic Diffusion-Transfer Process," Ger. Offen. 2,004,798 (1970), 49 pp.; Chem. Abstr. 73, 406 (1970).

208. Beavers, D. J., E. S. Perry, and W. J. Staudenmayer (Eastman Kodak Co.), "Protein-free, Image-receiving Layer with Silver Precipitating Agent and a Colloidal Binding Agent for the Diffusion Transfer Process," Ger. Offen. 2,004,799 (1970), 33 pp.; Chem. Abstr. 73, 406 (1970).

209. Kodak N. V., "Development Nuclei for the Silver-Salt Diffusion Process," Neth. Appl. 6,607,192 (1966), 7 pp.; Chem. Abstr. 66, 7574 (1967).

210. McGuckin, H. G. (Eastman Kodak Co.), "Photographic Antihalation Compositions and Layers," French Patent 1,480,864 (1967), 4 pp.; Chem. Abstr. 67, 11,445 (1967).

211. Yudelson, J. S., and H. J. Gysling (Eastman Kodak Co.), "Palladium Salt Photographic Process," French Patent 1,573,592 (1969), 23 pp.; Chem. Abstr. 72, 389 (1970).

212. House, H. O., Modern Synthetic Reactions, W. A. Benjamin, Inc., New York, N. Y., 1965.

213. Bond, G. C., and E. J. Sercombe, "Supported Platinum Metal Catalysts. Their Selection and Methods of Use in Industrial Processes," Platinum Metals Rev. 9 (3), 74-79 (1965).

214. Rylander, P. N., "Double Bond Isomerization as a Product-Controlling Factor in Hydrogenation over Platinum-Group Metals," in Platinum-Group Metals and Compounds, Advances in Chemistry Series, No. 98, American Chemical Society, Washington, D. C., 1971, pp. 150-162.

215. Anonymous, "Hydrogen Peroxide," Hydrocarbon Process, 40 (11), 255 (1961).

216. Bailar, J. C., Jr., "The Homogeneous Hydrogenation of Soybean Oil Methyl Ester," Platinum Metals Rev., 15 (1), 2-8 (1971).

217. Anonymous, "Demand for Caprolactam Spurs Expansions," Chem. Eng. News, 51 (30), 6-7 (1973).

218. Prengle, W., Jr., and N. Barona, "Make Petrochemicals by Liquid Phase Oxidation. Part 2. Kinetics, Mass Transfer and Reactor Design," Hydrocarbon Process., 49 (11), 159-175 (1970).

219. Anonymous, "Acetaldehyde from Ethylene," Hydrocarbon Process., 40 (11), 206 (1961).

220. Johnson, B. F. G., "The Platinum Metals in Organic Synthesis. Organometallic Complexes as Preparative Reagents," Platinum Metals Rev., 15 (2), 60-67 (1971).

221. Manassen, J., "Homogeneous Catalysis with Macromolecular Ligands," Platinum Metals Rev., 15 (4), 142-143 (1971).

222. Anonymous, "Details Emerge on Palladium Catalysis," Chem. Eng. News, 47 (17), 48-50 (1969).

223. Anonymous, Internal Report 1973 [on emissions from smelters and the Iron Ore Recovery Plant], International Nickel Company, Sudbury, Ontario, 1973.

224. Goldberg, R. N., and L. G. Hepler, "Thermochemistry and Oxidation Potentials of the Platinum-Group Metals and Their Compounds," Chem. Rev., 68, 229-252 (1968).

225. Anonymous, "Physical Properties of the Platinum Metals," Platinum Metals Rev., 16 (2), 59 (1972).

226. Hartley, F. R., The Chemistry of Platinum and Palladium with Particular Reference to Complexes of the Elements, John Wiley and Sons, New York, N. Y., 1973.

227. Lisk, D. J., "Trace Metals in Soils, Plants, and Animals" in Advances in Agronomy, Vol. 24, Academic Press, New York, N. Y., 1972, pp. 267-325.

228. Mellor, J. W., "Chapter 71. Palladium" in Inorganic and Theoretical Chemistry, Vol. 15, Longmans, Green and Company, London, 1947, pp. 592-685.

229. Rose, A., and E. Rose, The Condensed Chemical Dictionary, 7th ed. Reinhold Publishing Corp., New York, N. Y., 1966.

230. Kleinberg, J., W. J. Argersinger, Jr., and E. Griswold, Inorganic Chemistry, D. C. Heath and Company, Boston, Mass., 1960.

231. Hodgman, C. D., R. C. Weast, and S. M. Selby, Eds., Handbook of Chemistry and Physics, 40th ed., Chemical Rubber Publishing Company, Cleveland, Ohio, 1958.

232. Cotton, F., and G. Wilkinson, Advanced Inorganic Chemistry. A Comprehensive Text, Interscience Publishers, a Division of John Wiley & Sons, U. S. A., 1962.

233. Muller, O., and R. Roy, "Synthesis and Crystal Chemistry of Some New Complex Palladium Oxides" in Platinum Group Metals and Compounds, Advances in Chemistry Series No. 98, American Chemical Society, Washington, D. C., 1971, pp. 28-38.

234. Walsh, T. J., and E. A. Hausman, "The Platinum Metals" in Treatise on Analytical Chemistry. Part II. Analytical Chemistry of the Elements Vol. 8, I. M. Kolthoff and P. J. Elving, Eds., Interscience Publishers John Wiley and Sons, New York, N. Y., 1963.

235. Weast, R. C., Ed., Handbook of Chemistry and Physics, 50th ed., Chemical Rubber Publishing Co., Cleveland, Ohio, 1969-1970.

236. Heathcote, L. A., "Bright Palladium Plating. Production of Pore-free Deposits at Room Temperature," Platinum Metals Rev., **9** (3), 80-82 (1965).

237. Anonymous, Materials and Services for Research, Pressure Chem. Co., Pittsburgh, Pa., 1972.

238. Schmoyer, L. F., Engineering Manager, Pressure Chemical Co., personal communication, 1973.

239. Dickens, P. G., R. Heckingbottom, and J. W. Linnett, "Oxidation of Metals and Alloys. Part 2. Oxidation of Metals by Atomic and Molecular Oxygen," Trans. Faraday Soc., **65** (8), 2235-2247 (1969).

240. Nabivanets, B. I., L. V. Kalabina, and L. N. Kudritskaya, "State of Palladium(II) and Platinum(IV) in Inorganic Acid Solutions," Izv. Sib. Otd. Akad. Nauk SSSR, Ser. Khim. Nauk, (4), 51-53 (1970); Chem. Abstr. 74, 312 (1971).

241. Defense Documentation Center, Trace Metal Effects, A Report Bibliography, Search Control No. 083144, Defense Supply Agency, Alexandria, Va., 1972.

242. Otsuka, S., Y. Tatsuno, and K. Ataka, "Univalent Palladium Complexes," J. Am. Chem. Soc., 93, 6705-6706 (1971).

243. Beamish, F. E., The Analytical Chemistry of the Noble Metals, Pergamon Press, London, 1966.

244. Anonymous(A.S.D.), "Palladium-Silicon Alloys," Platinum Metals Rev., 16 (2), 49 (1972).

245. Guiot, J. M., "Interaction of Oxygen with Palladium Powder," J. Appl. Physics, 39 (7), 3509-3511 (1968).

246. Rytvin, E. I., and Ulybysheva, L. P., "Mechanism of Volatilization and Oxidation of Platinum, Palladium, and Their Alloys," Izvestiya Akademii Nauk SSSR, Metally, (1), 247-252 (1969); Chem. Abstr. 70, 275 (1969).

247. Kubachewski, O., "Practical Aspects of Metallurgical Thermodynamics," Platinum Metals Rev., 15 (4), 134-140 (1971).

248. Bond, G. C., "The Organic Chemistry of Palladium" [Book Review], Platinum Metals Rev., 16 (2), 56 (1972).

249. Bond, G. C., "Platinum and Palladium Complexes of Unsaturated Hydrocarbons, A Comprehensive Review of the Literature," Platinum Metals Rev., 14 (2), 62-63 (1970).

250. Cassar, L., "Synthesis of Carboxylic Acids and Esters by Carbonylation Reations at Atmospheric Pressure Using Transition Metal Catalysts," Synthesis. Intern. J. Methods Synthetic Org. Chem., 1973 (9), 509-523 (1973).

251. Moritani, I., and Y. Fujiwara, "Aromatic Substitution of Olefins by Palladium Salts," Synthesis. Intern. J. Synthetic Org. Chem., 1973 (9), 524-533 (1973).

252. Bird, C. W., "Palladium-catalysed Reactions in Synthetic Organic Chemistry Chem. Ind. (London), (13), 520-522 (1972).

253. Trost, B. M., and T. J. Fullerton, "New Synthetic Reactions. Allylic Alkylation," J. Am. Chem. Soc., 95 (1), 292-294 (1973).

254. Hartley, F. R., "Platinum Metals in Organometallic Chemistry. The International Conference in Moscow," Platinum Metals Rev., 16 (1), 22 (1972).

255. Kuendig, P., M. Moskovits, and G. A. Ozin, "Matrix Isolation Laser Raman Spectroscopy. Raman and Infrared Spectral Study Including Raman Polarization Data for the Cocondensation Reaction of Palladium and Platinum Atoms with Carbon Monoxide. Spectroscopic and Structural Data for the New Palladium Tetracarbonyl and Platinum Tetracarbonyl," J. Mol. Struct. 14 (1), 137-144 (1972); Chem. Abstr., 78, 482 (1973).

256. Kuendig, E. P., M. Moskovits, and G. A. Ozin, "Intermediate Binary Carbonyls of Palladium $Pd(CO)_n$ Where n = 1-3. Preparation, Identification, and Diffusion Kinetics by Matrix Isolation Infrared Spectroscopy," Can. J. Chem., 50 (22), 3587-3593 (1972); Chem. Abstr., 78 (8), 594 (1973).

257. Aston, J. G., "Thermodynamic Studies on the Hydrogen-Palladium System," Engelhard Ind. Tech. Bull., 7 (1/2), 14-20 (1966).

258. Pinta, M., Detection and Determination of Trace Elements, Distributor Daniel Davey and Co., Inc., New York, N. Y., 1966.

259. Anonymous, "Sample Collection and Preservation Procedures for Neutron Activation Analysis" and "University of Missouri Neutron Activation Analysis Detection Limits," Environmental Trace Substances Center, University of Missouri, Columbia, Mo., [undated].

OSMIUM

Ivan C. Smith
Thomas L. Ferguson

SUMMARY

The annual sales of osmium in the United States are relatively small, about 2,000 to 3,000 troy ounces (1 troy ounce = 31.1 g) per year. However, it is estimated that an additional 4,000 to 5,000 oz are toll-refined for captive use by industry. The 2,000 to 3,000 oz sold annually probably represent the maximum quantity lost to the environment each year through man use. The high cost of osmium and fluctuating prices ($300 to $450 per troy ounce) have limited its usage to a relatively few specialty items which take advantage of either its high melting point (3050°C), its unique catalytic properties, or its hardness and corrosion resistance.

The chemical industry consumed almost two-thirds of the new osmium production in 1971. Other major users include the medical and dental industries and the electrical industry.

Osmium consumed by the chemical industry is used in research and development as evidenced by the large number of literature citations devoted to research on osmium compounds. A significant fraction of osmium is also consumed as a catalyst for steroid synthesis. No evidence of osmium contamination of steroids intended for human consumption has been reported, although this exists as a remote possibility. The majority of osmium used by this industry would apparently be purchased as the tetroxide.

Osmium consumed by the electrical industry is used primarily in electrical contacts, such as relays, or in incandescent lights because of its high melting point. However, osmium oxidizes at low temperatures (approximately 100°C) in air; consequently, devices using osmium are encapsulated to prevent formation of the volatile osmium tetroxide. A large fraction of the electrical devices which consume osmium are reprocessed to recover this expensive metal.

Osmium tetroxide is used by the medical industry for staining of tissue for histological examination and as a fixative for specimens to be examined by electron microscopy.

Osmium alloys have been used for such items as ballpoint pen tips, fountain pen nibs, cutting tools, and phonograph needles. The quantity used for each item is small and is widely distributed in the environment when disposed of. Such uses should pose no hazard to man or to the environment.

Osmium metal poses no recognized health hazard unless it is oxidized to the volatile tetroxide. The hazard of osmium tetroxide has been recognized since the time of the discovery of osmium. It is poisonous, irritating to the eyes and mucous membranes, and stains the skin and other tissues through reduction to the nontoxic or slightly toxic osmium dioxide or osmium metal.

The rapid and complete reduction of osmium tetroxide by organic materials should render this metal innocuous in the environment. Other chemical forms of osmium (VIII) generally hydrolyze to the tetroxide and would be detoxified by organic materials in the environment.

Strong oxidizing agents, e.g., hydrogen peroxide or hypochlorous acid, in an acidic medium are required to convert the dioxide to the tetroxide. There would be some opportunity for conversion to the tetroxide during the chlorination of wastewater in a water treatment plant; however, any tetroxide formed should rapidly revert to the dioxide through reaction with residual organic materials or when the effluent reaches the receiving stream.

The volatility of osmium tetroxide requires that it be used in well ventilated areas (fume hoods), and protective clothing must be worn when working with this compound. A significant vapor pressure of this compound exists over a water solution of the material, as evidenced by the ability to effect codistillation of osmium tetroxide with water. Caution is recommended when working with osmium-containing solutions.

Information gathered during this study indicates that only about 2,000 troy ounces of osmium are consumed in the United States annually. No indication of human, animal, or plant health hazards attributable to this metal or its compounds has been identified. The small quantities consumed by the users and their wide geographical distribution would make identification of any health hazards difficult.

We have estimated that an additional 1,000 to 3,000 troy ounces of osmium are being lost to the environment each year from refining of copper sulfide ores. This osmium would be lost as osmium tetroxide vapor or as osmium tetroxide absorbed on particulates lost during refining. This small quantity of the metal lost by the many copper refineries does not appear to pose an environmental hazard. However, sampling and analysis for osmium in the vicinity of such smelters should be conducted to substantiate this conclusion.

In general we conclude that: (1) the toxicity of osmium tetroxide is well recognized and poses no serious health hazard when handled by recommended procedures and with adequate caution; (2) no large localized loss of osmium sufficient to create an obvious environmental hazard was identified; (3) there are insufficient data in the literature on environmental contamination by osmium in the vicinity of potential sources such as copper refineries to identify a hazard; and (4) there are insufficient data to define the chronic toxicity of osmium compounds.

I. INTRODUCTION

This document summarizes available information on the natural occurrence, refining, processing, uses, and disposal of osmium and its compounds, as well as the effect of these activities on man and the environment. Osmium is rare in nature, quite expensive, and has achieved only limited commercial use. Existing markets for this metal are very competitive; consequently, suppliers and dealers are reluctant to disclose information on either uses or consumers. For these reasons only scattered data are available on the consumption of osmium and its compounds. The probable distribution pathways and ecological consequences of osmium presented in this report have been based primarily on subjective evaluations.

The findings of this study are presented in the following four major sections: Background; Mining and Processing; Consumers and Uses; and Environmental Effects.

Appended to this report is a brief review of the chemistry of osmium (Appendix A), a tabulation of the physical properties of osmium compounds (Appendix B), and an extended bibliography listing all documents and literature sources compiled on osmium.

II. BACKGROUND

Osmium was discovered by Smithson Tennant in 1804.[1] This element was first isolated from the black powder remaining after platinum had been recovered from platinum-bearing ores. Osmium occurs naturally in association with iridium and other platinum group metals. The most important source of this metal is as an alloy called osmiridium whose osmium content may vary considerably. The osmium content of osmiridium recovered as a by-product of gold mining in South Africa, for example, varies from 24.0 to 44.5%.[2] However, the most common composition contains about 35% osmium and 30% iridium. The rest of the alloy consists mainly of platinum, rhodium, and ruthenium. Osmiridium having the highest known osmium content was reported by Vernadsky[3] in 1914 to contain 80% osmium.

Levy and Picot,[4] during microscopic examination of various platiniferous concentrates, discovered a mineral whose optical properties were not described in the literature. Using the Castaing electron microprobe and chemical analysis, they showed that this mineral was pure osmium. No other reference to the discovery of pure osmium in nature has been found.

Although osmium is always found with the other five platinum-group metals, seldom is detailed information available on the osmium content of crude ore; instead statistics on osmium are often combined with those of ruthenium and iridium since none of these metals are produced in large volumes.

The six platinum-group metals commonly occur in placers as two intergrown alloys.[5/] One alloy has a high platinum content, a lower iridium content, and small contents of ruthenium, palladium, rhodium, and osmium. The second alloy consists mainly of iridium and osmium, with considerable ruthenium, less rhodium and platinum and very small quantities, if any, of palladium.

The platinum-group metals that occur in bedrock lodes exist mainly as the minerals. These metals, together with gold, silver, nickel, copper, and certain other elements, exist as cations combined with arsenic, antimony, bismuth, sulfur, tellurium, oxygen, and other anions. Small amounts of native alloy may occur in these bedrock ores.

A third type of lode is the deposits of the platinum-group metals which occur in peridotites and perknites, commonly in dunite and serpentenite. Platinum metals in these lodes occur as native alloys which are concentrated in lenticular masses of chromite. Such deposits have yielded no major lodes of economic value though some small ore bodies of extremely high grade have been found. Prolonged erosion and alluvial concentration of these lodes produce stream and beach placers such as those in the Russian Urals, Colombia, Alaska, and as far as known, all other platinum placers of the world.

The principal sources of platinum metals in the world are the Bushveld Complex of South Africa, the Sudbury, Ores of Canada, and the Urals of the USSR. Minor sources are placer deposits in Alaska, Colombia, Ethiopia, Japan, Australia, and Sierra Leone. Since 1934 Alaska has ranked fifth in world production of platinum. The Witwatersrand District of the Bushveld Complex of South Africa, however, is the world's greatest producer of osmiridium, the principal source of osmium.

Platinum metals have been found in 22 states of the United States, but only Alaska is a major producer. The principal sources in Alaska are the Goodnews Bay District, Western Alaska, and a copper lode on Kasaan Peninsula, Southeastern Alaska, which was worked several years but is now closed. A small fraction (19%) of the domestic production of osmium came from the Goodnews Bay mining operation in 1970. In California and Oregon, small amounts of platinum but apparently no osmium are recovered as a by-product of gold placer mining. Several ounces of osmium are recovered as a by-product of the copper and nickel ore processing industry. About 81%

of the domestic production of osmium came from these latter two sources in 1970. Small gold-platinum and copper-platinum lodes have been mined in past years in the Rocky Mountain states, generally without a profit.

The composition of an unidentified sample received by the Bureau of Mines in 1945 from Discovery Claim, Solomon River (Goodnews Bay District) contained 11.00% osmium, 14.48% platinum, 71.22% iridium, and the remainder ruthenium and rhodium. The dross of platinum metals from the Goodnews Bay District contained a mean concentration of 2.56% osmium, 73.56% platinum, and 13.16% iridium. Table I, a table taken from Geological Survey Professional Paper No. 630,[5/] shows ore analyses from 1936 to 1967 for samples taken from the Goodnews Bay District.

TABLE I

PERCENTAGES OF PLATINUM METALS, GOLD, AND IMPURITIES, GOODNEWS BAY DISTRICT

(Based on data from Goodnews Bay Mining Company)

Year	Pt	Ir	Os	Ru	Rh	Pd	Au	Impurities
1936	68.39	12.72	3.24	0.24	1.46	0.23	0.45	13.27
1937	64.95	17.28	3.29	0.29	1.85	0.25	1.08	11.01
1938	72.19	11.24	2.24	0.17	0.99	0.29	1.69	11.19
1939	71.54	12.26	2.57	0.20	1.16	0.31	1.63	10.33
1940	71.77	12.34	2.56	0.19	1.16	0.32	1.80	9.86
1941	72.44	11.05	2.22	0.20	1.14	0.30	2.01	10.63
1942	72.50	10.37	2.14	0.16	1.31	0.36	2.73	10.43
1943	74.68	9.39	1.75	0.13	1.21	0.36	2.19	10.29
1944	74.67	9.65	1.82	0.14	1.21	0.37	2.13	10.01
1945	73.09	10.30	2.07	0.16	1.31	0.34	2.16	10.57
1946	76.24	7.61	1.42	0.10	1.12	0.39	2.91	10.21
1947	77.25	5.83	0.94	0.07	1.01	0.38	3.73	10.79
1948	77.47	6.20	1.01	0.08	1.04	0.39	2.83	10.98
1949	76.86	7.07	1.22	0.10	1.14	0.38	2.54	10.69
1950	75.46	9.13	1.59	0.14	1.23	0.35	1.64	10.46
1951	75.26	9.33	1.76	0.14	1.28	0.35	1.43	10.45
1952	73.23	11.10	2.15	0.18	1.30	0.33	1.32	10.39
1953	71.57	12.19	2.45	0.19	1.21	0.31	1.46	10.62
1954	73.31	10.91	2.02	0.17	1.19	0.34	1.37	10.69
1955	74.39	9.83	1.76	0.16	1.31	0.38	1.69	10.48
1956	76.19	8.27	1.10	0.10	1.16	0.37	1.57	11.24
1957	75.39	8.26	1.42	0.11	1.12	0.38	1.70	11.62
1958	75.03	7.96	1.25	0.11	0.98	0.33	1.70	12.64
1959	75.22	7.88	1.40	0.11	0.95	0.32	1.98	12.14
1960	75.64	7.85	1.34	0.11	0.89	0.36	1.54	12.27
1961	76.19	7.42	1.28	0.10	0.96	0.37	1.68	12.00
1962	76.16	7.22	1.25	0.10	0.96	0.37	1.82	12.12
1963	71.83	10.18	1.93	0.15	0.98	0.33	2.52	12.08
1964	69.42	11.92	2.38	0.19	1.00	0.29	3.11	11.69
1965	69.80	12.26	2.41	0.19	1.01	0.29	3.34	11.60
1966	67.62	12.89	2.56	0.19	1.04	0.28	3.85	11.57
1967	67.51	12.60	2.59	0.18	0.98	0.27	4.17	11.70
Weighted Means	73.62	9.94	1.89	0.15	1.15	0.34	2.06	10.85

III. MINING AND PROCESSING

In 1970, about one-third of the free world output of the platinum-group metals, which includes osmium, was produced by International Nickel Company of Canada, Ltd. (INCO),[6] from nickel-copper lodes of the Sudbury District. Crude residues containing the platinum-group metals are produced from the Canadian ore and shipped to INCO's Platinum Metals Refinery in Acton, London, England, where the platinum metals are separated and recovered in the pure form. Much of the osmium is returned to Canada, exported to the United States, and sold to Englehard Industries of Newark, New Jersey. No figures on the total amount of osmium recovered each year from this mine have been found.

The Witwatersrand District of South Africa is the world's largest producer of osmiridium, the principal source of osmium. Rustenburg Platinum Mines, Ltd., produces platinum-group metals from this ore, which is mined chiefly for platinum, and contributes almost two-thirds of the free world output of platinum-group metals. The bulk of the Rustenburg production goes to Johnson, Matthey and Company, Ltd., in Great Britain for processing. Much of this metal is eventually sold in the United States by subsidiary companies, Matthey Bishop, Inc., and Johnson, Matthey and Company, Inc.

Englehard Industries also refines concentrates and platinum-bearing matte from the Brakesprite Mine in the Republic of South Africa.

As previously mentioned, a small quantity of osmium is produced from the United States mining operations. The major fraction of the domestically produced osmium is recovered as a by-product of copper refining in Maryland, New Jersey, Texas, Utah, and Washington.[7] Some osmium is also recovered from a placer platinum deposit at Goodnews Bay, Alaska, and refined by Matthey Bishop, Inc., in Malvern, Pennsylvania.

A. Mining and Refining of Osmium

The method of mining employed to recover osmium containing ore is dependent on the ore body. Alluvial deposits are commonly recovered by dragline dredging. The alluvial deposits being mined include the Goodnews deposit in Alaska, gold placers in the Ural Mountains of the USSR, and the Colombia deposits. Ores from these deposits are generally concentrated by gravity separation. This same technique is used in South Africa to concentrate osmiridium.

Copper and nickel ores, which contain platinum metals, are commonly mined by open pit methods. No attempt is made to separate the platinum metals from the major component of the ores. Most of the U.S. domestic production of platinum metals is recovered as a by-product of the copper industry. It is estimated that 1 oz of the platinum metals is recovered from 35 tons of copper produced; however, no serious effort is made to recover the precious metals in high yield.

The platinum-group metals in the Sudbury District of Canada are mixed with the sulfide ores of copper and nickel. Concentrates of the copper and nickel sulfides are obtained by magnetic and flotation techniques. The nickel concentrate is roasted with a flux, melted into a matte, and cast into anodes for electrolytic refining from which the precious metal concentrate is recovered.

An extractive metallurgy procedure used by International Nickel Company of Canada for separation and recovery of platinum metals and to recover osmium from the Sudbury ores is typical of ore-refining methods used to recover this metal.[8/] The metallic fraction from the matte separation process contains all six platinum-group metals. The platinum metals remain in the anode slime when the metallic concentrates are electrolytically refined. In conventional treatment for case metal removal the anode slimes are often sulfated at 600°F to remove residual nickel and copper.

After extracting with water to recover the copper and nickel, the concentrate that remains contains the precious metals. To recover the precious metals, the concentrate is calcined in air at 1500° to 1800°F to eliminate sulfur, selenium, and arsenic impurities. Unless precautions are taken to recover osmium, any osmium that survived the 500° to 600°F sulfating reaction (described above) would probably be lost in the calcining operation.

To retain the osmium during sulfation, the sulfation temperature is reduced to 400°F.[9/] Under this condition, less than 5% of the osmium is lost. To recover osmium, the calcining operation has been modified and provisions made to recover it during this operation.[9/] Gradually raising the temperature to 1500° to 1700°F during this calcining gave satisfactory osmium elimination.

The osmium tetroxide in the off-gases is collected in a basic scrubbing solution. Typical recovery of osmium is 85%. Osmium is recovered as osmium metal sponge by partially neutralizing the alkaline scrubbing solution with sulfuric acid to pH 8 and gassing with sulfur dioxide to pH 6 to precipitate a complex sodium-osmium sulfite; treating the crude precipitate with sulfuric acid at 240°F to evolve sulfur dioxide and then with an oxidizing agent that volatilizes only osmium tetroxide (and ruthenium tetroxide if present); absorbing the distillate vapors in 20% sodium hydroxide solution; adding methanol to precipitate sodium ruthenate and reduce the perosmate to the osmate; filtering the alkaline osmate solution; treating the sodium osmate solution with excess saturated potassium hydroxide to precipitate potassium osmate; filtering, washing, and drying the salt; and treating with hydrochloric acid at 250°F under hydrogen pressure 350 psig. The sponge is pyrophoric if exposed to air, oxidizing to osmium tetroxide. This danger is eliminated by drying at 200°F in a hydrogen stream followed by annealing in a similar atmosphere at 1700°F.

Our calculations indicate that about 80% of the osmium in the anode slime is apparently recovered in this process. However, Mineral Facts and Problems[6/] states that 90% of the platinum metals are recovered from Sudbury ores. No information can be found on the percent of the osmium recovered from the crude ore.

Milling and beneficiation of platinum-bearing nickel ores from the Republic of South Africa consist essentially of gravity concentration, flotation, and smelting which produces a high-grade table concentrate called "metallics" suitable for direct chemical refining. The second product, the nickel-copper matte, is subsequently smelted and refined at the Johnson-Matthey Plant at Brimsdown, England. At Brimsdown, the process of extracting the platinum metals consists of enriching the platinum-nickel matte to about 65% platinum metals and then treating the enriched product with acids to separate the individual platinum-group metals followed by final refining. Residues containing the platinum-group metals are recovered from (1) the smelting operations where the original matte is broken down to yield metallic nickel and copper anodes, (2) from electrorefining operations where these anodes are dissolved electrolytically to produce pure nickel and copper and an anode slime containing precious metals, and (3) from chemical operations for the separation and refining of the individual metals.

Mining of crude platinum in placer deposits furnishes only a small part of total production. The mining and processing techniques for recovering crude platinum from placers are similar to those used for recovering gold. In the gold ores of Africa, osmiridium is recovered by gravity concentration and refined by methods described above.

The sources that account for the domestic U.S. production of osmium can be divided into five categories: new metal derived from crude platinum ores, osmium recovered as a by-product of gold and copper mining, metal from the processing of crude platinum obtained from foreign sources, Metal recovered from secondary sources such as scrap, and the osmium recovered from scrap and virgin materials from captive use by an industry (toll refined). The U.S. Bureau of Mines Minerals Yearbook[10/] statistics on osmium produced from these sources are shown in Table II. The amounts of new osmium derived from domestic sources and from foreign sources are shown in Table III. Although Minerals Yearbook has not reported osmium

TABLE II

NEW OSMIUM RECOVERED BY REFINERS IN THE U.S. BY SOURCE[10]
(Troy Ounces)

Year	From Domestic Sources Crude Platinum, Gold and Copper Refining	From Foreign Crude Platinum
1962	95	5
1963	189	0
1964	366	149
1965	315	884
1966	219	1,314
1967	151	--
1968	95	--
1969	135	--
1970	149	--
Total	1,714	2,352

TABLE III

OSMIUM PRODUCED IN THE U.S.[10]
(Troy Ounces)

Year	New Metal	Recovery from Secondary Sources	Toll Refined	Total U.S. Production	Sold to Industries in U.S.	Refiner, Importer and Dealers Stockpile
1962	100	99[a]		199	1,125	2,762
1963	189	273		462	1,056	1,531
1964	515	928		1,443	1,379	1,936
1965	1,199	763		1,962	1,634	1,502
1966	1,533	728		2,261	1,836	2,559
1967	151	2,377		2,528	1,823	2,802
1968	95	672	2,920	3,687	1,612	2,402
1969	135	208	2,197	2,540	1,472	2,873
1970	149	121	958	1,228	1,707	1,868
1971	--	352	4,196	--	2,126	--

a/ These sources were not reported separately until 1968.

recovered from foreign crude platinum, information on imports of osmium, iridium, and osmiridium unwrought or partially worked are shown in Table IV. The amount of osmium derived from this foreign crude material cannot be determined. It is not known whether this material is refined for domestic use or refined for export. There are obvious voids in the available data which prevent making a mass balance of osmium produced and consumed in this country. Available data on toll-refined osmium indicate that a large fraction of the osmium used in this country (50 to 70%) is controlled and used captively. The amount of osmium toll-refined each year is probably only a fraction of that held by individual industries.

TABLE IV

OSMIUM, IRIDIUM AND OSMIRIDIUM IMPORTED UNWROUGHT OR PARTLY WORKED[a/]

Country of Origin	Imports for Consumption (troy ounces) 1968[b/]	1969[c/]	1970[d/]	1971[e/]	1972 (January-November)[f/]
Canada	2,000	2,497	2,200	3,250	9,340
United Kindgom	11,977	7,169	5,531	21,611	32,152
Netherlands	25	--	1,000	--	345
West Germany	--	--	--	--	138
Republic of South Africa	3,448	487	--	3,000	3,398
Ireland	6	--	--	--	--
Switzerland	5	--	100	--	--
Norway	--	38	--	--	--
Japan	--	16	--	--	--
Australia	--	--	3	--	--
Total	17,461	10,207	8,834	27,861	45,373

a/ U.S. Department of Commerce, Bureau of the Census data on imports of commodity by country (FT 135 series).
b/ December 1968.
c/ December 1969.
d/ December 1970.
e/ December 1971.
f/ November 1972.

B. Recovery for Reprocessing

Various methods have been developed for recovering osmium from waste. The two major sources of osmium waste reprocessed in-house include catalysts (OsO_4) and fixative solutions (OsO_4) used in electron microscopy or for tissue staining. Scrap metal is simply returned to one of the refineries for reprocessing.

James Harkema[11/] of the Upjohn Company was issued a patent in June 1971 on a process for the recovery of osmium tetroxide used as a catalyst in steroid synthesis. It was reported that osmium can be recovered from a steroid reaction mixture in reusable form and in 90 to 100% yield by treating the reaction mixture with thiourea under aqueous acidic conditions to form a thiourea complex of osmium, which is separated from the reaction mixture as an aqueous solution. The osmium thiourea complex is then oxidized with hydrogen peroxide to osmium tetroxide, which is recovered in reusable form by standard methods; for example, by extraction with a water-immiscible organic solvent or by distillation. This process is reported to provide an effective method for removing osmium from the steroid reaction mixture, thereby preventing contamination of the steroid products with the highly toxic osmium tetroxide. Osmium tetroxide is reportedly a commonly used catalyst by many pharmaceutical houses for steroid production. The FDA and other government regulatory agencies do not analyze for such rare contaminants in pharmaceuticals, and it is not known whether osmium is a contaminant of steroids of the type synthesized with osmium tetroxide as a catalyst.

Jacobs and Leggitt[12/] reported recovering more than 80% of the osmium, as osmium tetroxide, from used fixative solutions. It was noted that these fixative solutions commonly contain excess ferrous sulfate which is added routinely to reduce the osmium tetroxide to osmium dioxide to make it safe for disposal. This process for recovering osmium consisted of adding concentrated hydrochloric acid and hydrogen peroxide to the tetroxide in water solution and recovery of the tetroxide by codistillation with water.

Schlatter, et al.[13/] reported a similar method for regeneration of used osmium tetroxide fixative solutions. Their method also employed hydrogen peroxide to oxidize osmium to the tetroxide followed by codistillation of the newly formed tetroxide and water. Osmium metal waste is simply shipped to a refinery for reprocessing.

Englehard Industries, a major producer of osmium, estimates that approximately 2,000 troy ounces of osmium were produced in 1971 in the U.S.[14/] This represents a significant increase over 1970 production if the figure is accurate. The Bureau of Mines estimated that 4,169 troy ounces of osmium were reclaimed by consumers for captive use in 1971.[7/]

Over the 9-year period from 1962 through 1970, approximately 934 lb were marketed in this country or about 100 lb/year.

C. Osmium Losses During Refining

The majority of osmium produced is recovered either as osmium tetroxide or metallic osmium. When the metallic form is to be produced, the tetroxide is precipitated as osmyltetraamine chloride $[OsO_2(NH_3)_4Cl_2]$ or as potassium osmate. The osmium metal is then formed by ignition of the former compound in hydrogen[8/] or by treating potassium osmate with hydrochloric acid and hydrogen.[9/] If osmium tetroxide is the desired product, the precipitation and ignition steps are omitted, and osmium tetroxide is distilled directly from the process liquor.

The magnitude of losses of osmium during mining and refining is difficult to estimate. Where recovery of platinum group metals is an integral part of the refining operation, the percent recovery of all of the metals except osmium is believed to be quite high. Osmium, on the other hand, is easily lost during smelting or roasting of concentrated ores because of the ease of oxidation of the metal at elevated temperatures and the volatility of the oxidation product, OsO_4. Internation Nickel claims a recovery efficiency of 90% for precious metals, osmium excluded, from their Sudbury ore.

No attempt is made to recover platinum-group metals from some ores, either because the platinum metals are present in too small quantities, or because their presence is not recognized. For example, the Salt Creek Mine at the northwestern extremity of the Kasaan Peninsula in Alaska produced a low-grade copper ore and small amounts of gold and silver as a by-product. In 1917, the owners found that the ore contained platinum metals which thereafter became the principal product of the mine.

Platinum metals are recovered from gold and copper in refineries. The source of the copper ore is usually from lode-mined ore, while the gold ores and bullion may come from either lode or placer mines. It is difficult to estimate how much platinum group metals, particularly osmium, is lost to the environment from these sources since it is very dependent on the recovery process used. Table II shows that 1,714 troy ounces of osmium were recovered from these sources between 1962 and 1970.

Data cannot be found on osmium and other platinum-group metals lost to the environment from ores not processed for these metals. It can generally be assumed, however, that the platinum-group metal content of these ores must be too low to justify the cost of the additional processing.

IV. CONSUMERS AND USAGE

Information has been collected from a variety of sources in an effort to identify osmium consumers, principal uses of osmium and its compounds, and loss to the environment from these sources.

Osmium has found only limited use in relatively expensive end-products because of its high cost, fluctuating price, and short supply. The only commercial-product forms of osmium produced in any volume are osmium metal and osmium tetroxide.

The Bureau of Mines[10] statistics indicate that the chemical industry is the major consumer of osmium in this country. In 1969 and 1970, the chemical industry purchased more than half of the total amount of osmium sold in the U.S.; in 1971, it purchased almost three-fourths of the osmium sold. Osmium consumed by this industry is apparently used for a variety of purposes, generally in small individual quantities. The medical and dental industry is the next greatest user. These industries (chemical, medical and dental) combined consumed greater than 95% of the total osmium production. The Baker Division of Englehard Industries,[15] which is their dental division, states, however, that no osmium is sold in the U.S. for usage in dental products. Thus, the medical industry apparently accounts for nearly all the osmium consumed by that category of industries.

The biggest use of osmium is as a catalyst for steroid production. The Upjohn Company purchases about 4,000 g (130 troy ounces) annually for this purpose.[16] The G. D. Searle and Company has purchased about 2,000 g of osmium per year for several years for use as a catalyst. This company recovers all spent catalyst and ships it to a refiner for reprocessing. Of the 2,000 g purchased, about half is recovered from the spent material. The major losses apparently occur at the refinery during the recovery operation. Merck at one time also used osmium tetroxide as a steroid dehydrogenation catalyst, but now uses a new synthesis route that does not employ osmium. No other pharmaceutical houses have been identified that use osmium catalysts.

Osmium metal was used in alloys for fountain pen nibs and ball-point pen tips until about 1969, when pen tip manufacturers changed to a ruthenium-platinum alloy because of the scarcity of osmium and fluctuating prices.[17] Osmium is still used for watch and clock bearings, engraving tools, and needles for phonographs.

It is estimated that perhaps 200 troy ounces are consumed yearly by the electrical industry[18] for contact points on reed switches. At the present time the use is still in the exploratory stage. If it proves feasible, use could develop into a substantial market.

A method for the electrodeposition of osmium was reported by L. Greenspan.[19] Osmium deposited by this process is thick and bright and plates at high cathode efficiencies. This, or a similar process, would be used to fabricate electrical contacts.

Osmium compounds at one time were used in fingerprint detection, although not for taking fingerprints. It is no longer used for this purpose.

The other major application appears to be the use of the tetroxide for histological staining of tissue samples for electron microscopy. This use apparently accounts for the major portion of the quantity consumed by the medical and dental industries. Mallinckrodt, a major distributer of this compound, sells an estimated 400 g of osmium tetroxide per year for this purpose.

Literature citations indicate that an osmium compound is being tested as a catalyst for electrochemical devices such as fuel cells. However, no evidence of its use for this purpose has been found.

A survey of users, producers, and distributors of osmium and osmium compounds yielded insufficient data to develop a mass balance for the osmium consumed annually in the U.S. Some large purchasers of osmium were identified, however, that account for about 25% of the annual osmium sales. In some instances, the uses made of these large purchases could not be determined. Table V contains a list of the largest sales of osmium. These sales occur annually.

TABLE V

LARGEST OSMIUM PURCHASES

Purchaser	Chemical Form	Use	Quantity Purchased (oz)
The Upjohn Company	OsO_4	Catalyst	130
G. D. Searle and Company	OsO_4	Catalyst	65
Unknown	Os metal	Electrical switches	200
Sandia Corporation	Os metal (?)	Unknown	10-50
Washington University	OsO_4	Tissue staining	12
U.S. Government[a]	OsO_4	Unknown	100-200

a/ The federal government purchases several large lots per year. It is not purchased through GSA.

The best assessment that can be made of osmium uses and consumption based on information obtained from producers, distributors, users, and a survey of the technical and patent literature and of trade journals, is tabulated below in the order of decreasing importance.

Consumer	Uses	Quantity (%)
Academic and other research laboratories	Chemical synthesis Catalysts Metallurgical Nuclear Physics Biological	45
Medical laboratories Medical research	Tissue staining for electron microscopy Chemotherapy	35
Chemical industry	Catalysts Steroids Polymers Hydrogenation	10
Electrical industry	Reed switches Light filaments Cathode tubes	5
Others	Mechanical pivots Bearings Phonograph needles Engraving tools	5

A. Research Uses

The major consumers of osmium appear to be academic and research laboratories. A survey of Chemical Abstracts indicates that over 150 new osmium compounds are synthesized and studied annually. Although no accurate estimate can be made of the amount of osmium used for this purpose, it appears to be significant. Many citations also appear on the use of osmium as hydrogenation, oxidation, and hydroxylation catalysts. Although investigated extensively as a catalyst, osmium has apparently found only limited use for this purpose. The only known industrial use of osmium as a catalyst is the synthesis of steroids. The tetroxide is also used as a reagent for analyzing polystyrene-butadiene block polymers.

Other literature citations that imply research-oriented uses of osmium include development of analytical methods, recovery of osmium from solutions, alloys of osmium, electroplating, use of osmium as nuclear targets, diffusion of osmium in metals, electron emission of osmium surfaces, spectral studies of osmium compounds, and many others. In addition, academic and other research laboratories consume osmium for histological staining of tissue and for electron microscopy studies. It must be concluded that a significant fraction of the 2,000 oz of osmium sold to U.S. consumers annually is consumed in small quantities in a large number of laboratories for the above purposes.

A large fraction of the osmium used in the synthesis and characterization of new compounds is retained on laboratory shelves and may only be discarded after several years. The osmium used for other research purposes (catalysts, development of analytical methods, histological fixative and stain, etc.) is largely discarded down the drain as it is used. Unless recommended disposal methods are developed and enforced, the majority of these materials will be discarded by flushing down the drain or as solid waste.

B. Medical Uses of Osmium

Considerable quantities of osmium tetroxide are used in medical research and in routine clinical laboratory testing. The major use of this material is for histological examination of tissue specimens. Osmium tetroxide has long been recognized as an excellent fixative, applicable to a wide range of tissue types, which causes the least disturbance of cell structure.[20-25] This property has resulted in the wide use of osmium tetroxide to prepare thin sections of tissue for electron microscopic examination.

Each of the osmium suppliers identified medical uses as major consumers of osmium compounds. Only a few, however, can be identified. The Medical School at Washington University, St. Louis, Missouri, purchases about 16 oz of osmium tetroxide or 12 oz of osmium metal per year. Other major consumers identified include the Veterans Administration Hospital in Milwaukee, Rochester University, and Kent University. A limited survey of other medical schools and hospitals indicates that purchases of 1 to 15 oz of osmium tetroxide per year are common for those institutions that have electron microscopes. The University of Kansas Medical School uses about 12 oz/year, and the University of Missouri Medical School uses 1 to 2 oz of osmium tetroxide per year.

One supplier has reported that several thousand grams of osmium tetroxide are purchased annually by the U.S. Government. This material is not purchased through GSA, making difficult determination of the quantities purchased and the uses. The seller of this osmium tetroxide is reported to have the best material for tissue staining; consequently, it is assumed that the majority of this material is used as a tissue fixative and stain.

Nearly all the osmium used for histological study is disposed of by flushing down the drain even by the larger consumers. Although recovery of osmium from used fixative solutions is relatively simple, it is not practiced widely because of the small amounts consumed annually in an individual laboratory and the hazards associated with recovery. Two similar methods are described in the literature for recovering this osmium (see Section III-B).

The only chemotherapeutic use of osmium identified is for treatment of arthritis. Osmic acid injections were found by von Reis and Swensson[26] to be one of the most effective treatments for rheumatoid arthritis. They showed that the entire interior synovial layer coagulated as a result of the injections. Severe pain and a fever reaction were caused by the osmic acid injection. Other investigators have modified this treatment to combine osmic acid, a local anesthetic, and corticosteroids to reduce pain and other side reactions.[27,28]

C. Chemical Industry

A substantial fraction of the osmium consumed by the chemical industry is apparently used for research purposes. Again, many of the research areas described in Section IV-A are being investigated in chemical industry research laboratories.

To our knowledge, the only chemical-industrial uses of osmium involve osmic acid as a catalyst for steroid synthesis and as an oxidation catalyst for determining the polystyrene in a butadiene-styrene copolymer.

The single biggest use of osmium is as a catalyst for steroid production. The chemical form of osmium used for this purpose is osmium tetroxide. It is used as a catalyst in the manufacture of such steroids as cortisone, hydrocortisone, prednisone, prednisolone, 2-α-methylprednisolone, 16-methylhydrocortisone, 6-α-fluoroprednisolone, acylates, and derivatives of these compounds. A large fraction of the osmium used for this purpose is reclaimed and used again. The Upjohn Company purchases approximately 4,000 g (130 troy ounces) to replenish its stock annually.[16] Merck at one time also used osmium tetroxide as a steroid dehydrogenation catalyst. It now uses a new synthesis route which does not employ osmium.

D. Electrical Industry

No large uses of osmium by the electrical industry have been identified. It was estimated by one of the major osmium suppliers that 200 oz of osmium has been used in the production of reed switches. The use of osmium for this purpose is still experimental, and only began to develop after a process was found for electroplating osmium.[19] Should this use grow, it is anticipated that osmium used for this purpose will be recycled with no substantial discard to the environment.

An osmium-tungsten alloy has been used for filaments for incandescent lamps. These filaments contain 3 to 30 wt % osmium. These filaments are strong up to 2200°C and are used for high-intensity lamps such as the more expensive motion picture projectors.

Because of the high work function and high melting point of osmium, the metal is of interest as a coating in vacuum tubes to suppress secondary electron emission from tungsten and molybdenum grids. No evidence of its use for this purpose has been found, however. A tungsten alloy containing 0.1 to 10% osmium has been patented for use as x-ray tube anode plates. A patent has also been issued on a cathode for electronic devices made of osmium and thorium. We have not been able to determine that significant quantities of osmium are used for this purpose.

The quantity is believed to be very small since Bureau of Mines statistics indicate that only 5 oz and 2 oz of osmium were consumed by the electronics industry in 1969 and 1970, respectively. Any large quantities used by this industry would likely be recycled.

E. Other Uses

Osmium has, over a number of years, been used for a variety of purposes. Most of these uses took advantage of the corrosion resistance and the hardness of the metal and its alloys.

Two alloys[29] of osmium (Os-86%, Ir-5%, Pt-10%, and B-1%; and Os-85%, Rh-5% and Pt-10%) were used as fountain pen nibs on the better pens until 1969, at which time pen manufacturers turned to a ruthenium-platinum alloy because of the scarcity of osmium and its fluctuating prices.[17] Osmium was also used for fingerprint detection; however, this use was discontinued because it caused dermatitis.

Osmium, or its alloys, is still used for pivots, bearings, phonograph needles, and engraving tools because of its hardness. The actual quantity used for this purpose could not be determined; however, it is small. It was noted by one manufacturer that 50,000 pivots could be fabricated from 1 oz of osmium. Osmium is one component of a complex catalyst employed in a hydrogen-chlorine fuel cell patented by Union Carbide.[30] Applications have been made for U.S. patents on modifications of electrodes for use in a fuel system of this type. To our knowledge, no commercial fuel-cell system employs this catalyst.

V. ENVIRONMENT EFFECTS

Osmium consumed as alloys poses no apparent health hazard. The osmium in each of the items is used in extremely small quantities. This highly corrosion-resistant metal should survive intact for many years when disposed of as solid waste. Incineration of the waste would convert it to the toxic osmium tetroxide; however, the quantities present in each unit of solid waste are too small to be of concern.

Osmium, being extremely rare, is not considered by the authors to be a serious environmental contaminant even though the toxicity of osmium tetroxide has been recognized almost from the time of the discovery of osmium. Following is a review of the sources of possible environmental contamination of osmium and an estimate of their hazards.

The metal and its natural and synthetic alloys pose no recognized health hazard. These compounds are inert to chemical attack. However, osmium metal begins to oxidize in air at relatively low temperatures (approximately 100°C) to form osmium tetroxide. The temperature required for such oxidation is well above any ambient temperature that the element or its alloys might encounter in nature.

Since osmium is found in nature primarily as the osmium-iridium alloy (osmiridium), there are no natural sources that should pose a hazard. Consequently, it must be concluded that any hazardous environmental contamination must be man-made.

Since we have found no analytical information on osmium contamination of the biosphere even in the vicinity of likely man-made sources, we can only estimate the sources and magnitude of man-made contamination and its influence on the biosphere.

The U.S. reserves of platinum metals are almost entirely in copper ores. The largest sources of osmium lost to the air would come from roasting and smelting of copper concentrates, which contain small quantities of platinum-group metals. In a roasting and smelting operation, osmium present in the ore could be converted to the volatile osmium tetroxide. Osmium tetroxide begins to sublime below 30°C and boils at 130°C; consequently, typical stack gas temperatures are sufficiently high to maintain osmium tetroxide in the vapor phase. Thus it appears that most osmium lost during copper refining would be emitted as a vapor; however, an unknown fraction of this compound could be absorbed on the particulates in the stack gas. Removal of OsO_4 from stack gases by any control process other than chemical scrubbing is highly unlikely.

The osmium that might be entering the environment from this source cannot be accurately estimated. No data can be found on the osmium content of crude ores. It has been estimated, however, that about 1 oz of platinum-group metals is recovered per 6,000 tons of ore.[6/] Bureau of Mines statistics show that about 258 million tons of copper ore were mined in 1970. Probably over 80% of this copper is mined as the sulfide ore.

The major steps in producing copper metal from low-grade sulfide ore are beneficiation, roasting, converting and refining.

Flotation is commonly used to concentrate the ore after it has been crushed, ground, and classified. No information is available on how much of the platinum metals might be lost during this operation.

Roasting is conducted in a reducing or at least a nonoxidizing atmosphere to produce a copper matte, which consists of a mixture of Cu_2S and FeS. The composition of this matte may contain from 15 to 50% copper. Precious metals in the concentrate dissolve in this matte. Production of a high-grade matte can result in poor recovery of precious metals. Some osmium loss could occur during this operation, probably as a consequence of inadequate recovery rather than by oxidation and volatilization even though the ore is subjected to temperatures of 1800°F in this operation.

Converting is the final stage in the smelting process and consists of passing a thin stream of air through the molten matte to oxidize the FeS to eliminate the sulfur as SO_2 and to form a ferrous slag and blister copper. The temperature of the molten matte is 1800°F. It is during this operation that the greatest osmium loss would occur. In studies conducted by Illis et al.[9/] on calcination of precious metal concentrates, to eliminate S, Se, etc., it was found that more than 95% of the osmium is volatilized at 1700°F.

Although less osmium should be volatilized during the converting of copper matte to blister copper than during the calcining of precious metal concentrates, substantial losses likely occur during this operation.

None of the gaseous effluent control devices presently in use would recover vaporized osmium tetroxide.

The final step in copper refining involves fire-refining of the blister copper, which includes an oxidation operation. Substantial losses of osmium could occur during this process. After fire-refining, the metal is cast into anodes suitable for electrolytic refining. It is the anode slimes remaining after electrolysis that are processed for platinum metals.

It must be concluded that the small amount of osmium recovered annually as a by-product of copper refining is only a fraction of that present in the original ore, probably less than 5 to 10%. Thus, 1,000 to 3,000 oz of osmium is probably being lost to the environment each year from copper smelters.

It is speculated that a modest fraction of the osmium sold annually is lost to the environment. The mode of disposal is dependent on the type of usage.

The largest fraction of osmium sold annually is consumed by the chemical industry. The large number of literature citations found annually on new compounds of osmium indicates that a substantial fraction of the osmium sold to the chemical industry is used for research purposes in academic and other research laboratories. Total number of ounces consumed for this purpose has not yet been determined. However, it would appear that a large fraction of the osmium is retained in individual laboratories in the forms of new chemical compounds of that metal.

The second largest use is for tissue-staining. The majority of osmium consumed for this purpose is apparently discarded to the environment in wastewater.

Osmium consumed in manufacturing, generally as catalysts, is commonly recycled and reused many times. Unavoidable losses during production account for most of the annual sales to the chemical manufacturing industry. All manufacturing organizations we have contacted indicated that they reprocess and reclaim osmium. Waste osmium lost by these users is commonly flushed down the drain, generally with no pretreatment to control toxicity. Because of the high organic content of most wastewater, it is probable that almost all of the osmium, particularly osmium tetroxide, is reduced either to osmium dioxide or to osmium metal soon after it reaches the wastewater stream.

The fate of osmium in wastewater is not known. Osmium disposed of by pouring down the drain will pass through wastewater treatment plants where the majority of the metal will be removed in the sludge. Soluble forms remaining in the water could be converted to osmium tetroxide during the chlorination operation and volatilized into the air. However, the majority of the soluble osmium should pass through the chlorination process unaffected. The small percentage present as osmium tetroxide will react rapidly and completely with residual organic matter in the receiving stream and be converted to osmium metal or to osmium dioxide and settle out in the sediment of the water course. The extremely low levels of osmium expected to be found in wastewater streams should pose no hazards to man or his environment.

The small amount of osmium metal disposed of as solid waste is resistant to chemical attack, and because of its wide dispersal into the environment should pose no health hazard. The only situation which could create a hazard consists of incineration of wastes containing osmium metal. Under incineration conditions, the osmium will be converted to volatile osmium tetroxide. However, the quantities normally involved in an incineration operation are probably too small to pose a hazard.

VI. HEALTH HAZARDS

Metallic osmium is considered inert.[31/] However, when heated in air or in a finely divided form at room temperature, it oxidizes to form osmium tetroxide. This compound is volatile, highly irritant, and toxic. In absence of data to the contrary, it should be assumed that other soluble compounds of osmium, at least those in the higher valence state, are probably also toxic. Finely divided or sponge osmium is pyrophoric and should be handled with care. Osmium dioxide, which is believed to be a product of the reduction reaction when osmium tetroxide reacts with organic materials, is also considered to be inert.

Inhalation of osmium tetroxide results in severe irritation of the mucous membranes of the nose, throat, and bronchi and can produce headaches. One fatal case has been reported, with death described as being due to bronchitis and bronchial pneumonia. Kidney injury was also found and has been observed in animal experiments with osmium tetroxide. Osmium compounds have a caustic action on skin causing eczema and dermatitis. Irritation of the eyes is ordinarily the first symptom from exposure to low concentrations of osmium tetroxide. Lacrimation, a gritty feeling in the eyes, and appearance of halos around lights are frequently reported. In most cases recovery occurs after a few days. Instillation of one drop of 1% osmic acid into the eye of a rabbit results in severe eye damage. There are apparently no published data on orally ingested osmium. However, soluble osmium compounds would be expected to be highly toxic by this route.

Individuals working in the vicinity of osmium tetroxide should take precautions to prevent inhalation, skin contact, eye contact, or ingestion. It is recommended that those individuals use respiratory protective devices and wear protective clothing and goggles to prevent contact with this compound. It is also recommended that osmium tetroxide only be used in well-ventilated areas, preferably in fume hoods.

Osmium tetroxide is a powerful oxidizing agent, and accidental combination with reducing agents should be avoided. As a catalytic agent, it will cause hydrogen-oxygen mixtures to explode at slightly elevated temperatures.

APPENDIX A

CHEMISTRY OF OSMIUM AND ITS COMPOUNDS

A detailed review of the chemistry of osmium and its compounds is not within the scope of this report; however, a general review is appropriate to show how losses can occur and how this metal can react after it enters the environment.

Finely divided osmium or sponge osmium, which has a high surface area, oxidizes in air to form the poisonous osmium tetroxide. The high-surface area material is so reactive that it will explode in air. At least part of this reactivity may be due to hydrogen adsorbed during manufacture.

The finely divided metal is attacked by oxidizing acids such as fuming nitric acid, *aqua regia*, and alkaline hypochlorite solutions. The compacted metal is not attacked by nonoxidizing acid but is slightly attacked by *aqua regia* and nitric acid.

Osmium exhibits oxidation states of I, II, III, IV, V, VI, VII, and VIII. An oxidation state of IX has been reported; however, reexamination of the results and a better understanding of the chemistry of osmium cast doubt on the existence of the reported compound.

OSMIUM(VIII)

Osmium(VIII) oxide is the most important compound of osmium and together with the free metal is the starting point for nearly all the chemistry of osmium. The tetroxide is readily formed by heating the metal in air or by distilling an osmium-containing compound from nitric acid or other solutions containing an oxidizing agent such as hypochlorous acid or hydrogen peroxide.

An aqueous solution of osmium tetroxide is neutral. Its first acid dissociation constant is about 8×10^{-13}. The compound is soluble in water, alcohol, and ether.

An alkaline solution of osmium tetroxide (perosmate) is reduced by alcohol or nitrite to osmate(VI). Potassium osmate(VI) dihydrate can be precipitated from solution by the addition of excess potassium hydroxide.

Reactions of osmium tetroxide are presented in Figure A-1.[32] Other compounds of Os(VIII) that have been prepared include the oxyfluorides $[OsO_3F_2]_n$, $K[OsO_3F_3]$, and $Cs_2[OsO_4F_2]$ (no unsubstituted fluorides); potassium perosmate $K_2[OsO_4(OH)_2]$; nitride complexes such as $K[OsO_3N]$; the imide complex $Me_3CN:OsO_3$; the amine complex $OsO_4 \cdot NH_3$; $OsO_4 \cdot py$ and other pyridine complexes; and 1,4-dichlorobutadiene adducts.

OSMIUM(VII)

Compounds of Os(VII) that have been characterized include $K_3[OsO_5]$, $Na_2[OsO_6]$ (and the analogous lithium and barium salts), OsF_7, and $OsOF_5$.

OSMIUM(VI)

Compounds typical of this oxidation state, which have been characterized, include OsP_2; OsF_6 (the only hexahalide); the oxyhalo species $[OsOCl_4]_n$, $K_2[OsO_2Cl_4]$, and $Cs[OsOCl_6]$; $Na_4[OsO_5]$; nitrido complexes such as $K_2[Os\overline{N}Cl_5]$ and others substituting H_2O for labile halogen ligands; $OsO_2(NH_3)_4Cl_2$; $[Os(en\text{-}H)_4]I_2$ where en-H = ethylenediamine minus one proton; $Ospy_2O_3$; sulfito complexes; osmium esters formed in hydroxylations of olefins; cyano complexes $K_2[OsO_2(CN)_4]$ and *trans*-$K[OsN(H_2O)(CN)_4]$; and other osmyl complexes most of which have the general formula $M_2^I[OsO_2X_mY_{4-m}]$ where M = K, Na, or NH_4 and X and Y are the same or different and include OH, Cl, Br, OMe, CN, salicylate, NO_3, NO_2, or SO_3. For these complexes m = 1-4. When X or Y = oxalato, m = 1 or 2. OsO_3 exists in oxygen at 800° to 1500°.

OSMIUM(V)

Os(V) is in an extremely unfavorable oxidation state. The few compounds reported include OsF_5, hexafluoroosmates, $Sr_2Cr^{III}[OsO_6]$, $K[OsF_6]$, $NO[OsF_6]$, $SeOsF_9$, possibly Os_2O_5 and Os_2S_5, Os_2NCl_7 (which is probably Os_2NCl_5), and $[Os(en\text{-}H)_3en]I_2$ (which is unlikely to contain Os(V) since it is diamagnetic).

OSMIUM(IV)

Binary compounds of Os(IV) include OsF_4, $OsCl_4$, $OsBr_4$, OsO_2 (and its mono- and dihydrates), OsS_2, $OsSe_2$, and $OsTe_2$. The relatively few

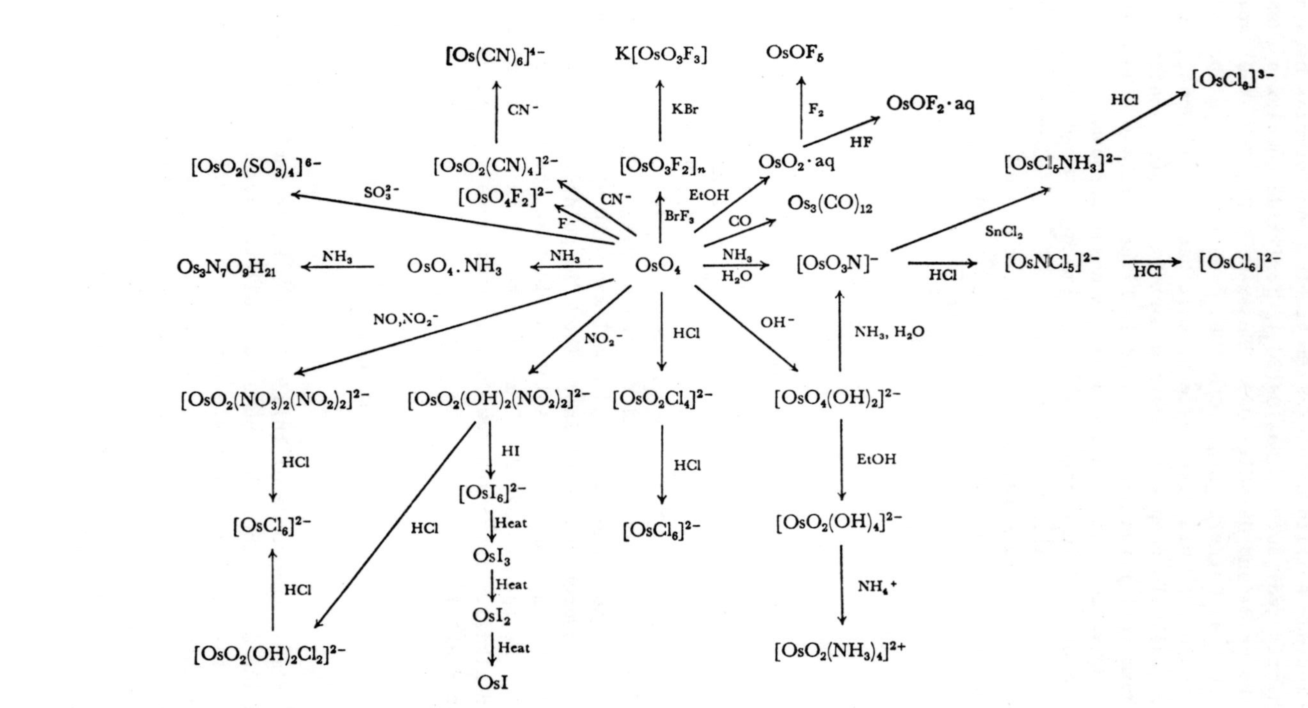

Figure A-1 - Reactions of Osmium Tetroxide[32]

complexes include sulfito complexes (most of whose formulas have not been fully authenticated) such as $Na_3[Os(SO_3)_6]$; nitrido complexes $[OsN(NH_3)_8X_2]X_3$ where X = Br or Cl and Os_2NCl_5; $OsO_2 \cdot 2NH_3 \cdot H_2O$; $OsCl_4 \cdot 2NH_3 \cdot H_2O$; stable arsine complexes such as $Os(AsPh)_3Br_4$; $[CpOs(OH)]^+$ where Cp = cyclopentadiene; and complexes having bidentate heterocyclic bases as ligands, e.g., 2,2'-dipyridyl. The halo complexes are the most important. Reactions of the hexachloroosmate(IV) ion are presented in Figure A-2.[32/] The white potassium, cesium, sodium, ammonium, and barium hexafluoroosmate(IV) salts are stable in water or dilute acid. Hexaiodoosmate(IV) is unstable, more water-soluble, and less resistant to hydrolysis. Of the sodium, ammonium, rubidium, cesium and silver hexabromoosmates(IV), only the sodium salt is soluble in water. The mixed hexahaloosmates $[OsCl_nX_{6-n}]^{2-}$ where $\underline{n}$ = 1-5 and X = Br or I and the oxychloride $(NH_4)_4[Os_2OCl_{10}]$ are known.

OSMIUM(III)

Compounds representative of this oxidation state are Os_2O_3 (dubious existence), hexachloro- and, perhaps, hexabromoosmate(III) salts that are unstable in aqueous solutions; OsX_3 where X = Cl, Br, or I; a few amine complexes such as $[Os(NH_3)_6]Br_3$ and $K_2[Os(NH_3)Cl_5]$; nitro complexes such as $Na_2Os(NO_2)_5 \cdot 2H_2O$ that are fairly stable in cold aqueous solutions; phosphine, arsine, and stibine complexes; numerous bidentate-heterocyclic-ligand complexes; and complexes with acetylacetonato ion, thiourea, selenourea, and pyridine. There is polarographic evidence for Os(III) species in reduced OsO_4 solutions in base and acid and for cyano complexes. The species $Os(OH)_6^{3-}$ may be present in the brown solution of "Os_2O_3" in alkali. There are few Os(III) complexes with ligands having oxygen donor atoms.

OSMIUM(II)

Simple compounds of this oxidation state include $OsSO_3$ and the only dihalide OsI_2. There are numerous well-characterized complexes of Os(II) with heterocyclic bidentate ligands; SO_3; pyridine, NO; phosphine, arsines, stibines; PF_3; CN^- (e.g., $K_4[Os(CN)_6]$, which is colorless and fairly water-soluble and whose anion is very stable); CO; H; alkyl groups; aryl groups; and olefins. Osmocene and substituted osmocenes have been characterized. The band spectrum of OsO has been measured, but there is little evidence that the solids reported are OsO.

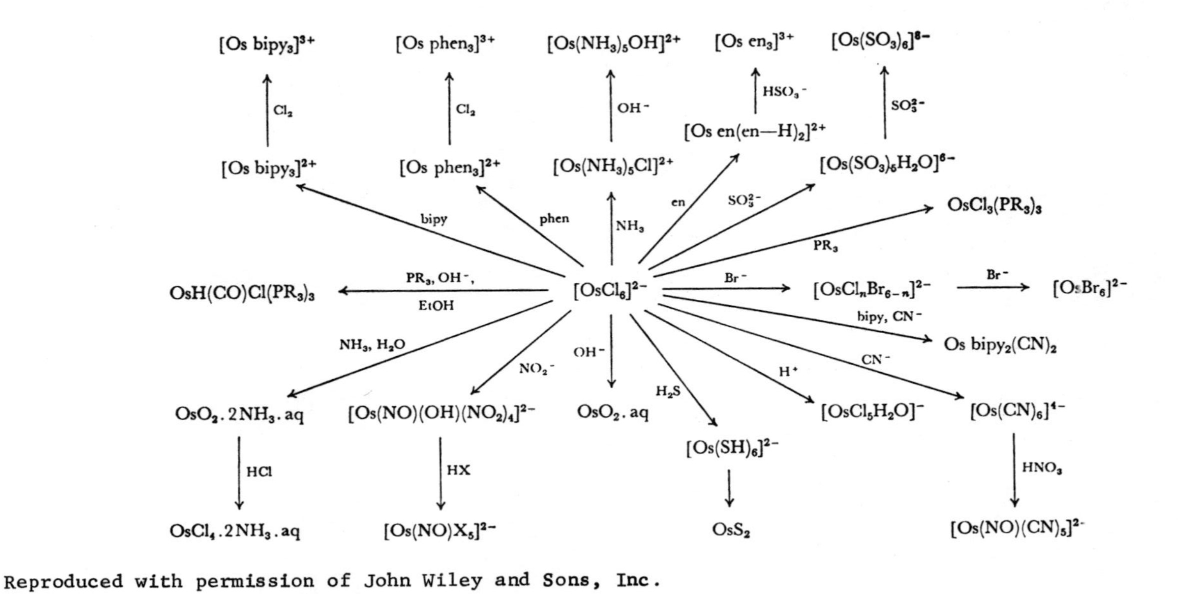

Figure A-2 - Reactions of $[OsCl_6]^{2-}$ 32/

OSMIUM(I)

Os(I) compounds that have been characterized are $[Os(NH_3)_6]Br$; carbonyl complexes such as the stable $[Os(CO)_4X]_2$, $[CpOs(CO)_2]_2$, and $CpOs(CO)_2Br$ where Cp = π-C_5H_5; and OsI, the only monohalide.

OSMIUM(0)

The complexes of Os(0) include the very unstable $Os(NH_3)_6$; $Os(PF_3)_5$; the carbonyls $Os(CO)_5$, $Os_3(CO)_{12}$, $Os(CO)_4H_2$, and $Os(CO)_3(PPh_3)_2$; and arene olefin complexes such as (π-C_6H_6)$Os(C_6H_8)$, π-benzene cyclohexa-1, 3-diene osmium(0).

Although the hydrolytic stability of many of these compounds has not been reported, it appears that most of the halides and oxyhalides would revert to the oxides in the environment. The higher oxidation states would likely be reduced to the dioxide and perhaps even to the metallic forms through reaction with organic materials or other reducing agents in nature.

Rates of hydrolysis of classes of these compounds should be determined to provide guidance in selecting suitable methods of disposal.

OSMIUM COMPOUNDS

Formula	Oxidation State	Boiling Point (°C)	Melting Point (°C)	Solubility	Physical Form	Behavior in Air, Water, Etc.
$Me_3CN:OsO_3$	VIII		Decomp. 112[a]		Yellow crystals[a]	
$K_2[OsO_4(OH)_2]$, Potassium perosmate	VIII			Very sol.[a] H_2O	Deep red crystals[a]	Easily reduced to K osmate[a]
$K[OsO_3N]$, Potassium osmiamate	VIII		Explodes 180[b]	Sl. sol. H_2O (Na salt very sol.)[a]	Yellow cryst. powder[b]	Other alkali metal, Ba, Zn, Ag, Tl, Hg, and NH_4 salts made; only the Na salt is very H_2O-sol.[a]
OsO_4	VIII	131.2[a]	40.6[a]	Sp. sol. H_2O,[a] very sol. CCl_4, alc., ether[c]	Only one form. Pale yellow solid[a]	Vapor pressure: log p = 10,000/(4.57 T + 5.49)[c]; (10 mm at 26°)[d]; 100 μg/m^3 causes severe conjunctival irritation.[a] Threshold limit value 0.002 mg/m^3 in air.[d] Pungent chlorine-like odor.[d]
OsS_4[?[a]]	VIII		Decomp.[e]	Insol. H_2O, sol. dil. acid and alkali[e]	Brown-black crystals[d,e]	
$K_3[OsO_5]$	VII			Decomp. H_2O[a]	Black[a]	Disproportionates in H_2O to OsO_4 and K osmate.[a]
OsF_7	VII				Yellow solid[a]	
$OsOF_5$	VII		59.8[a]		Green crystals[a]	
$K_2[OsO_2(OH)_4]$	VI			Sol. H_2O[b] insol. in alc., ether[c]	Violet crystals[b]	Poisonous. Slowly decomp. in aq. solns with formation of OsO_4.[c]
$K_2[OsNCl_5]$	VI			Very sol. H_2O, insol. org. solvents[a]	Reddish purple crystals[a]	Decomp. in dil. solns.[f] Labile to H_2O substitution.[a]
$[OsO_2(NH_3)_4]Cl_2$	VI			Sl. sol. H_2O[a]	Orange yellow solid[a]	
$(OsOF_4)_n$	VI		Sublimes 90		Golden yellow crystals[a]	
$(OsOCl_4)_n$	VI	200[a]	32[a]		Dark brown crystals[a]	
OsF_6	VI	45.9[a] (47)[g]	32.1[a,g]	Decomp. in H_2O[e]	Green deliquescent crystals[b,e] Volatile yellow solid[f]	Decomp. rapidly in H_2O to OsO_4, HF, and $[OsF_6]^-$.[g] Decomp. under UV irradn. to OsF_5.[g]

Formula	Oxidation State	Boiling Point (°C)	Melting Point (°C)	Solubility	Physical Form	Behavior in Air, Water, Etc.
OsP_2	VI			-	Nonvolatile gray-black solid	Inert to acid and alk. solns. Decomp. to elements at 1000° in vacuo.[f]
OsF_5	V	225.9[a]	70		Blue[a] (green[g]) crystals	
$(NH_4)_2OsCl_6$	IV			Sol. H_2O and alc.[c]	Orange rhombic prisms[c]	
K_2OsCl_6	IV			Sol. H_2O, sparingly sol. alc.[c]	Dark red to almost black octahedral crystals[c]	
OsO_2	IV		Decomp. 650[e] Decomp. $\geq$ 400-60[h]	Insol. H_2O and acid[e] Sol. HCl[a]	Two forms:[a] cubic or hexagonal red brown crystals[e] Brown or black solid[b] or copper-red octahedral or hexagonal crystals[h]	Reacts with O_2 at high temperatures to give OsO_4. May kindle or detonate depending on preparation.[h]
$OsO_2 \cdot 2H_2O$	IV			Insol. H_2O[b]	Black[b] amorphous[h] solid	May be pyrophoric oxidizing to OsO_4. Liable to detonate. Adsorbs foreign ions in its commonly colloidal state.[g]
OsS_2	IV		Decomp.[e]	Insol. H_2O, alkali[e]	Cubic black crystals[d,e]	Probably the only stable sulfide.[a]
$OsTe_2$	IV		~ 600[e]	Insol. in alkali and acid except HNO_3.[e]	Gray-black crystals[e]	
OsF_4	IV		230[a]	Decomp. in H_2O[e]	Brown powder.[e] Black volatile solid.[b,f] Yellow solid.[a,g]	Gives a clear yellow soln. in water from which OsO_2 is rapidly deposited.[f]
$OsCl_4$	IV	450[a]	Sublimes[e]	Sl. sol., decomp. in H_2O.[e] Sol. H_2O.[a] A red form insol. H_2O.	Red cryst. powder.[c] Red brown needles.[e] Black solid.[b]	Yellow water soln. gives osmium oxides (ultimately hydrated dioxide) + HCl on standing.[a,g]
$OsBr_4$	IV		Decomp. in vacuo 350[a]	Water and concd. acid insol.[a] A H_2O-sol. form.[a]	Black crystals.[a]	Hydrolyzed very slowly by alkalis. Thermal decompn. gives $OsBr_3$.[a]

Formula	Oxidation State	Boiling Point (°C)	Melting Point (°C)	Solubility	Physical Form	Behavior in Air, Water, Etc.
OsI_4[?a/]	IV			Very sol. H_2O[g]	Violet-black solid, hygroscopic, metallic lustre.[e,f]	Cold water soln. is stable; hot water evolves HI.[g]
$K_3[OsCl_6]$	III			Very sol. H_2O and alc.[a]	Red crystals.[a]	Unstable solns. give hydrated oxides on standing.[a]
$Na_2Os(NO_2)_5 \cdot 2H_2O$	III			Sol. H_2O[b]	Orange-yellow crystals.[b]	
Os_2O_3[?a/]	III		Decomp.[e]	Insol. H_2O[e]	Dark brown solid.[e]	
$OsCl_3$	III		Decomp.[g] 560-600[e] (Decomp. 450)[a] Sublimes ~ 350[g]	Very sol. cold H_2O; sl. sol. ether; acid-, org., and H_2O-insol. form[a,e]	Cubic brown crystals.[e] Brown-black hygroscopic powder.[b] Dark gray powder.[a]	Stable in H_2O and alc. even on boiling.[b,g] Resistant to mild reducing agents.[g] Gives weakly acidic solns. in H_2O. Freshly prepared solutions give no free Cl^-.[f]
$OsCl_3 \cdot 3H_2O$	III		Decomp.[e]	Very sol. cold H_2O; sol. alc.[e]	Dark green crystals.[e]	Probably $[Os(OH)Cl_3]_n$ [a]
$OsBr_3$	III			Insol. H_2O, acids and org. solvents[a]	Dark green powder[a]	Decomp. at high temp. to Os.[a]
OsI_3	III				Black amorphous solid[a]	
$Os(NO_2)_3$	III			Sol. H_2O[b]	Deep brown solid.[b]	Air stable.[a]
$K_2[Os(NO)Cl_5]$	II[a]			Sol. H_2O[b]	Dark red to black solid.[b]	
$(\pi\text{-}C_5H_5)_2Os$, Osmocene	II		229[a]		Colorless crystals.[a]	
OsO	II			Insol. H_2O and acid	Black[e]; dark gray amorphous powder[a]	
$OsCl_2$[?][a]	II		Decomp.[e]	Insol. H_2O; sol. ether, HNO_3; sl. sol. alkali[e]	Dark brown deliquescent solid[e]	Decomp. slightly in hot water.[e] Very inert.[g]
$Os(CO)_3Cl_2$	II		Decomp. 280[e]	Insol. H_2O[e]	Colorless prisms.[e]	
OsI_2	II				Black amorphous solid.[a]	
$OsSO_3$	II		Decomp.[e]	Insol. cold H_2O; sol. dil. HCl, alkali[e]	Blue-black crystals[d,e]	

Formula	Oxidation State	Boiling Point (°C)	Melting Point (°C)	Solubility	Physical Form	Behavior in Air, Water, Etc.
Os	0	5500[c]	2700[c]		Bluish-white lustrous metal. Cubic structure.[c]	Vapor pressure at m.p. 13.5 μ. Slowly oxidizes in air even at ordinary temperatures when finely divided. Oxidizes to OsO_4 in air > 200°.[c]
$Os(CO)_5$	0		-15	Sol. most org. solvents[2]	Colorless monomeric liq.[a]	
$Os_3(CO)_{12}$	0		224[a]		Yellow crystals.[a]	

a/ Griffith, W. P., The Chemistry of the Rarer Platinum Metals, (Os, Ru, Ir, and Rh), John Wiley and Sons (New York) (1967).

b/ Kirk-Othmer Encyclopedia of Chemical Technology, 2nd Edition, Anthony Standen, Executive Editor, Interscience Publishers, a Division of John Wiley and Sons, Inc., New York (London) (1963), pp. 874-875.

c/ Merck Index, 7th ed., Merck and Company, Inc., Rahway, New Jersey, 1960.

d/ Sax, Irving N., Dangerous Properties of Industrial Material, 2nd Edition, Reinhold Pub. Corp (New York) (1963).

e/ Handbook of Chemistry and Physics, Chemical Rubber Publishing Company, Cleveland, Ohio (1969-1970).

f/ Kleinberg, Jacob, Wm. J. Argersinger, Jr., and Ernest Griswold, Inorganic Chemistry, D.C. Heath and Company (Boston) (1960).

g/ Cotton, F. Albert, and G. Wilkinson, Advanced Inorganic Chemistry, Interscience Publishers, Division John Wiley and Sons (1962).

h/ Mellor, J. W., A Comprehensive Treatise on Inorganic and Theoretical Chemistry, Vol. XV, Longmans Green and Company, London (New York, Toronto) (1947). pp. 686-728.

REFERENCES

1. Tennant, S., Phil Trans., 94, 411 (1804).

2. Ball, J. E., and K. M. McBreen, "Platinum-Group Metals" in Minerals Yearbook 1954, Vol. I, Metals and Minerals (Except Fuels), Bureau of Mines, U.S. Department of the Interior, U.S. Government Printing Office (Washington, D.C.), pp. 915-931 (1958).

3. Vernadsky, V. I., Supplement to Handbuch der bestimmenden Mineralogie, (original publication Vienna, 1845), p. 725 (1914).

4. Levy, C., et al., Bull. Soc. Fr. Miner. Crist., 84, 312 (1961).

5. Mertie, J. B., Jr., "Economic Geology of the Platinum Metals," Geological Survey Professional Paper 630 (1969).

6. Ageton, R. W., and J. P. Ryan, "Platinum-Group Metals" in Minerals Facts and Problems, Bureau of Mines Bulletin 650, U.S. Department of the Interior, U.S. Government Printing Office (Washington, D.C.), pp. 653-669 (1970).

7. Mitko, F. C., U.S. Bureau of Mines, private communication, November 1972.

8. Hampel, C. A., Rare Metals Handbook, 2nd Edition, Reinhold Publishing Corporation (London) (1961).

9. Illis, A., B. J. Brandt, and A. Manson, Metallurgical Transactions, 1, 431 (1970).

10. Mitko, F. C., "Platinum-Group Metals" in Minerals Yearbook 1970, Vol. I, Metals, Minerals, and Fuels, Bureau of Mines, U.S. Department of the Interior, U.S. Government Printing Office Stock No. 2404-1126 (Washington, D.C), pp. 939-950 (1972).

11. Harkema, J., Process for the Recovery of Osmium Tetroxide, U.S. Patent 3,582,270, 1 June 1971.

12. Jacobs, G. F., and S. J. Liggett, Stain Technology, 46, 4, 207 (1971).

13. Schlatter, Ch., and I. Schlatter-Lanze, Journal of Microscopy, 94 (1), 85 (1971).

14. Carcillio, N., Englehard Industries, private communication.

15. Toliver, Baker Division of Englehard Industries, private communication, December 1972.

16. Schneider, A. W., Fine Chemical Division of the Upjohn Company, private communication, November 1972.

17. Hauck, J., W. A. Sheaffer Pen Company, private communication, February 1973.

18. Greenspan, L., Englehard Industries, private communications.

19. Greenspan, L., Journal of Plating, 59 (2), 137 (1972).

20. Hardy, W. B., J. Physiol., 24, 158 (1899).

21. Monkeberg, G., and A. Bethe, Arch. Mikr. Anat., 54, 135 (1899).

22. Strangeways, T. S. P., and R. C. Canti, Quart. J. Micr. Sci., 71, 1 (1927).

23. Porter, K. R., J. Exp. Med., 97, 727 (1953).

24. Frederic, J., Exp. Cell Res., 11, 18 (1956).

25. Wigglesworth, V. B., Proc. Roy. Soc., B147, 185 (1967).

26. von Reis, G., and A. Swensson, Acta Med. Scand. Suppl., 259, 27 (1951).

27. Berglöf, F. E., Acta Rheum. Scand., 5, 70 (1959).

28. Hurri, L., K. Sievers, and M. Oka, Acta Rheum. Scand., 9, 20 (1963).

29. Dennis, W. H., Metallurgy of the Nonferrous Metals, Sir Isaac Pitnam and Sons (London) (1954).

30. Kordesch, K. V., German Patent 1,271,797, 4 July 1968.

31. Anon., "Osmium and Its Compounds," American Industrial Hygiene Assoc. J., 29, 621 (1968).

32. Griffith, W. P., The Chemistry of the Rarer Platinum Metals, (Os, Ru, Ir, and Rh), John Wiley and Sons (New York) (1967).

EXTENDED OSMIUM BIBLIOGRAPHY

Ageton, R. W., and J. P. Ryan, Mineral Facts and Problems, Bureau of Mines Bulletin 650, U.S. Department of the Interior, U.S. Government Printing Office (Washington, D.C.) pp. 653-669 (1970).

American Metal Market, Metal Statistics 1972, Fairchild Publications, Inc. (New York) (1972).

Anonymous, 1972 Director of Chem. Producers-USA, Products Section, Chem. Information Service (SRI) (1972).

Anonymous, "Electrodeposition of Osmium," Platinum Metals Review, **16** (3), 90 (1972).

Anonymous, "Electroplating the Platinum Metals," Platinum Metals Review, **14** (3), 93, July 1970.

Anonymous, "Industrial Toxicology," Journal of Pharmacy and Pharmacology, **5**, 149-150 (1953).

Anonymous, "Osmium and Its Compounds," American Industrial Hyg. Assoc. J., **6** (29), pp. 621-623 (1968).

Anonymous, "Osmium Compounds," Kirk-Othmer Chem. Enc., Vol. 15, pp. 874-875 (1963).

Anonymous, "Physical Properties of the Platinum Metals," Platinum Metals Review, **16** (2), 59 (1972).

Anonymous, "Platinum Group Metals: Osmium," Kirk-Othmer Chem. Enc., Vol. 15, p. 858 (1963).

Anonymous, "Tidy Market for Shale," Chemical Week, **41**, 17 May 1972.

Appleby, A. J., "Electroplating of Osmium," J. Electrochem. Society, **117** (12), 1610 (1970).

Ball, J. E., and K. M. McBreen, "Platinum-Group Metals" in Minerals Yearbook 1954, Vol. I, Metals and Minerals (Except Fuels), Bureau of Mines, U.S. Department of the Interior, U.S. Government Printing Office (Washington, D.C.), pp. 915-931 (1958).

Beamish, F. E., The Analytical Chemistry of the Noble Metals, Pergamon Press, London (1966).

Berglöf, F. E., Acta Rheum. Scand., **5**, 70 (1959).

Bond, G. C., G. Webb, and P. B. Wells, "Hydrogenation of Olefins--Part 4, Reaction of n-Butene with Hydrogen Catalyzed by Alumina-Supported Ruthenium and Osmium," Trans, Faraday Soc., **64** (11), 3077-3085 (1968).

Bradford, C. W., "The Carbonyls of the Platinum Group Metals," Platinum Metals Review, **16** (2), 50 (1972).

Brunot, F. R., "The Toxicity of OsO_4 (Osmic Acid)," J. Ind. Hyg., **15** (1) 136-143 (1933).

Carter, F. E., "The Noble Metals Find Increasingly Wide Use in Industry," Materials and Methods, **28** (5), 55 (1948).

Charrin, V., "Platinum and the Platinum Group Metals," Genie Civil, **134**, 249 (1957).

Clark, Michael A., and Adolph G. Ackerman, "Osmium-Zinc Iodide Reactivity in Human Blood and Bone Marrow Cells," Anat. Rec., **170**, 81-96 (1970).

Clayton, Donald D., and William A. Fowler, "Abundances of Heavy Nuclides," Ann. Phys. (New York), **16**, 51-68 (1961).

Collan, Y., C. Servo, and I. Winblad, "An Acute Immune Response to Intra-Articular Injection of Osmium Tetroxide," ACTA Rheumatol. Scand., **17**, 236-242 (1971).

Committee on Threshold Limit Values, Amer. Conf. of Govt. Ind. Hygienists, Documentation of Threshold Limit Values (1971).

Costescu, L. M., and T. C. Hutchinson, "The Ecological Consequences of Soil Pollution by Metallic Dust from the Sudbury Smelters," Proc. Inst. Environ. Sciences, 18th Tech. Version, 540 (1972).

Cotton, F. Albert, and G. Wilkinson, Advanced Inorganic Chemistry, A Comprehensive Text, Interscience Publishers (USA) (1962).

Coughlin, J. P., "Contributions to Data on Theoretical Metallurgy XII," U.S. Bureau of Mines Bulletin No. 542 (1954).

Coughlin, J. P., "Contributions to Data on Theoretical Metallurgy XII," U.S. Bureau of Mines Bulletin No. 592 (1961).

Csanyi, L. J., "On the Catalytic Properties of OsO_4," Acta Chim. Hungary, 21, 39 (1959).

Dennis, W. H., Metallurgy of the Nonferrous Metals, Sir Isaac Putnam and Sons (London) (1954).

Environmental Protection Agency, Water Quality Criteria Data Book, Vol. 2, Inorg. Chem. Pollution for Freshwater. EPA 18010 DPV 07/71, Water Pollution Control Research Series, June 1971.

Faye, G. H., "Determination of Ruthenium and Osmium in Ore and Metallurgical Concentrates and in Osmiridium," Analytical Chemistry, 37 (6), 696 (1965).

Frederic, J., Exp. Cell. Res., 11, 18 (1956).

Gafafer, W. M., Occupational Diseases, U.S. Department of HEW, Public Health Service, PHS 1097 (1964).

Gilchrist, J. D., Extraction Metallurgy, Pergamon Press (1967).

Gilman, Henry (Editor), Organic Chemistry--An Advanced Treatise, Vol. 4, John Wiley and Sons, Inc. (London), p. 1180 (1953).

Goldberg, Robert N., and Loren G. Hepler, "Thermochemistry and Oxidation Potentials of the Platinum Group Metals and Their Compounds," Chemical Reviews, 68, 229-252 (1968).

Greenspan, Larry, "Electrodeposition of Osmium," Plating, 59 (2), 137-139 (1972).

Greenspan, Larry, "Electrodeposition of Osmium," Englehard Ind. Tech. Bulletin, 10 (2) 48 (1969).

Griffith, W. P., The Chemistry of the Rarer Platinum Metals, (Os, Ru, Ir, and Rh), John Wiley and Sons (New York) (1967).

Griffith, W. P., "Osmium and Its Compounds," Quarterly Rev., 19 (3), 254 (1965).

Gunstone, F. D., "Hydroxylation Methods III. Osmium Tetroxide," Advances in Organic Chemistry: Methods and Results, Interscience Publishers (New York), 1, 110-111 (1960).

Hair, M. L., and P. L. Robinson, "Complexes of Octavalent Ru and Os With Phosphorus Trifluoride," J. Chem. Soc., 106 (1958).

Hair, M. L., and P. L. Robinson, "The Reaction of Ru and Os Tetroxides with Ammonia," J. Chem. Soc., 2775 (1960).

Hampel, Clifford A., Rare Metals Handbook, Vol. 2, 2nd Edition, Reinhold Publishing Corporation (London), pp. 304-334 (1961).

Hanker, J. S., et al., "Coordination Polymers of Osmium: The Nature of Osmium Black," Science, 1737 (1967).

Hardy, W. B., J. Physiol., **24**, 158 (1899).

Harkema, James, Process for the Recovery of Osmium Tetroxide, U.S. Patent No. 3,582,270, 1 June 1971.

Hatch, Lewis F., "Homogeneous Catalysis for Liquid Phase Oxidation," Hydrocarbon Processing, **49** (3), 101 (1970).

Hawley, J. E., and Y. Rimsaite, "Platinum Metals in Some Canadian Uranium and Sulfide Ores," Amer. Mineralogist, **38**, 463 (1953).

Hayat, Arif M., "Principles and Techniques of Electron Microscopy: Biological Applications," Vol. 1, Van Nostrand Reinhold Corporation (New York, New York) (1970).

Holloway, J. H., "The Fluorine Compounds of the Platinum Metals," Platinum Metals Review, **16** (4), 118 (1972).

Howland, A. L., J. W. Peoples, and E. Sampson, The Stillwater Igneous Complex and Associated Occurrences of Nickel and Platinum Group Metals, Montana Bureau of Mines and Geol., Misc. Contribution No. 7, April 1936.

Hoyt, Charles D., and Patrick J. Ryan, "Platinum-Group Metals" in Minerals Yearbook, pp. 921-931 (1969).

Hunt, L. B., and F. M. Lever, "Availability of the Platinum Metals," Platinum Metals Review, **15** (4), 126-138 (1971).

Hunter, Donald, "Toxicology of Some Metals and Their Compounds Used in Industry," British Medical Bulletin, **7** (1-2), 5 (1950).

Hurri, L., K. Sievers, and M. Oka, Acta Rheum. Scand., **9**, 20 (1963).

Illis, A., B. J. Brandt, and A. Manson, "The Recovery of Osmium from Nickel Refinery Anode Slimes," Metallurgical Transactions, **1** (2), 431 (1970).

International Nickel Company, "The Platinum Group Metals in Industry," Brochure.

Jacobs, G. F., and S. J. Liggett, "An Oxidation Distillation Procedure for Reclaiming Osmium Tetroxide from Used Fixative Solutions," *Stain Technol.*, **46** (4), 207-208 (1971).

Jahn, C. A. H., "Platinum Metals--A Survey of Their Production, Properties and Engineering Uses," *Metal Industry*, p. 183, 5 March 1948.

Jahn, C. A. H., "Platinum Metals--A Survey of Their Production, Properties and Engineering Uses," *Metal Industry*, pp. 206-267, 12 March 1948.

Johnson, C., and R. H. Atkinson, "Refining Metals of the Platinum Group," *Ind. Chemist*, 223 (1937).

Johnson, F. F. G., "The Platinum Metals in Organic Synthesis," *Platinum Metals Review*, **15** (2), 60 (1971).

Johnstone, Rutherford T., and Seward E. Miller, *Occupational Diseases and Industrial Medicine*, W. B. Saunders Company (Philadelphia and London) (1960).

Kavanagh, J. M., and F. E. Beamish, "New Fire Assay for Osmium and Ruthenium," *Analytical Chemistry*, **32** (4), 490 (1960).

Keays, Reid R., and James H. Crocket, "A Study of Precious Metals in the Sudbury Nickel Irruptive Ores," *Economic Geology*, **65** (4), 438 (1970).

Kelly, K. K., "Contributions to the Data on Theoretical Metallurgy," U.S. Bureau of Mines Bulletin No. 601 (1962).

Kelly, K. K., and E. G. King, "Contributions to the Data on Theoretical Metallurgy," *U.S. Bureau of Mines Bulletin No. 592*, **593**, 73 (1961).

Kolthoff, I. M., T. S. Lee, and C. W. Carr, "Determination of Polystyrene in GR-s Rubber," *J. Polymer Sci.*, **1** (5), 429 (1946).

Kordesch, K. V., German Patent 1,271,797, 4 July 1968.

Kubota, Tokuo, Keiji Yoshida, and Fumihiko Watanabe, "The Configurations of the Products Obtained from Hydroxylation of Steroidal $\Delta^{1,4}$-Ketones with OsO_4," *Chemical and Pharmaceutical Bulletin*, **14** (12), 1426-1430 (1966).

Langley, R. C., "Osmium Films by Thermal Decomposition of Osmate Esters," Engelhard Ind. Tech. Bulletin, **10** (1), 16 (1969).

Levy, C., P. Picot, New Data on Iridium-Osmium Compounds. Existence of Native Osmium, Foreign Text Translated on Contract NASA TT F-13,758 by Translation Consultants, Ltd., Arlington, Virginia.

Liebenberg, W. R., "On the Origin of Uranium, Gold and Osmiridium in the Conglomerates of the Witwatersrand Goldfields," Neues Jahrbuck für Mineralogie Abhandlungen (Stuttgart), **94**, 831 (1960).

Lorinca, G., J. A. Isomaki, and J. Martio, "Changes in the Synovial Fluid Caused by Osmic Acid," Acta Rheumatol. Scand., **16**, 217-222 (1970).

Luning, David, "How to Buy Platinum-Group Metals and Chemicals," Chemical Engineering, 114-118 (1972).

Mancuso, T. F., E. J. Coulter, and E. J. McDonald, "Migration and Cancer Mortality Experience--A Study of Native and Southern Born Nonwhite Ohio Residents," in Trace Substances in Environmental Health - VI. Proceedings of the University of Missouri's 6th Annual Conference on Trace Substances in Environmental Health, Columbia, Missouri, 13-15 June 1971, D. D. Hemphill, Ed., Published by University of Missouri, Columbia, Missouri (1973).

McCleverty, J. A., "Fe, Ru, and Os Annual Survey Covering 1970," Organometallic Chem. Rev. B., **10**, 123-172 (1972).

McCleverty, J., "Fe, Ru, and Os, Annual Survey Covering 1968," Organometallic Chem. Rev. B., **5**, 419-474 (1969).

McDonald, D., A History of Platinum, Johnson Matthey and Company, Ltd. (London) (1960).

McLaughlin, A. I. G., R. Milton, and Kenneth M. A. Perry, "Toxic Manifestations of Osmium Tetroxide," Br. J. Ind. Med., **3**, 183 (1946).

Mertie, John B., Jr., "Economic Geology of the Platinum Metals," Geological Survey Professional Paper 630 (1969).

Miles, Nicholas, A., and Sidney Sussman, "The Hydroxylation of the Double Bond," J. Am. Chem. Soc., **58**, 1302 (1953).

Mitko, Francis C., "Platinum-Group Metals" in Minerals Yearbook 1970, Vol. I, Metals, Minerals, and Fuels, Bureau of Mines, U.S. Department of the Interior, U.S. Government Printing Office Stock No. 2404-1126 (Washington,D.C), pp. 939-950 (1972).

Monkeberg, G., and A. Bethe, Arch. Mikr. Anat., 54, 135 (1899).

Mottonen, M., K. Karkola, M. Pantio, and T. Nevalainen, "Histochemically Demonstrable Changes in Lactate and Succinate Dehydrogenases in Normal Rabbits' Synovial Membrane After Injection of Osmium," Acta Rheumatol. Scand., 17, 79-84 (1971).

Mottonen, M., M. Pantio, and T. Nevalainen, "Effects of Osmium Tetroxide on the Rabbit Knee Joint Normal Synovail Membrane," Acta Rheumatol. Scand., 16, 121-129 (1970).

Munro-Ashman, D., et al., "Contract Dermatitis from Palladium," Trans. St. John Hosp. Derm. Soc., 55, 196-197 (1969).

Nilas, N. A., and Sidney Sussman, "The Hydroxylation of the Double Bond," J. Am. Chem. Soc., 58, 1302 (1953).

Ogburn, S. C., Jr., "Some New Analytical Reactions of the Platinum Metals," J. Am. Chem. Soc., 48 (10) (1926).

Ogburn, S. C., Jr., "Qualitative Separation of the Platinum Metals," J. Am. Chem. Soc., 48, 2507-2512 (1926).

Oka, M., A. Rekonen, and A. Ruotsi, "The Fate and Distribution of Intra-Articularly Injected Osmium Tetroxide (Os-191)," Acta Rheumatol. Scand., 15, 35-42 (1969).

Parker, Raymond L., "Chapter D. Composition of the Earth's Crust" in Data of Geochemistry, 6th Edition, Michael Fleischer, Technical Editor, Geol. Survey Prof. Paper 440-D, U.S. Government Printing Office (Washington, D.C.), 19 pages (1967).

Peckner, Donald, "Precious Metals and Their Uses," Mater. Design Eng., 57 (6), 93-102 (1963).

Perry, K. M. A., "Diseases of the Lung Resulting from Occupational Dusts Other than Silica," Thorax, 2, 21 (1947).

Philpott, J. E., "Applications of the Noble Metals in the Chemical Industry," S. A. Chemical Processing, February-March 1969.

Pinta, Maurice, Detection and Determination of Trace Elements. Translated from French by Mirium Bivas and edited by IPST, Israel Program for Scientific Translations, Jerusalem, 1966 (Ann Arbor, Michigan: Ann Arbor Science Publishers, Inc., 1971).

Plummer, M. E. V., J. M. Kavanagh, J. C. Hole, and F. E. Beamish, "Copper, Nickel, and Iron Alloys for the Quantitative Recovery of the Platinum Metals in Ores and Concentrates," Transactions of the Metallurgical Society of AIME, **221**, 145 (1961).

Porter, K. R., J. Exp. Med., **97**, 727 (1953).

River, Charles, Assoc., Inc., Economic Analysis of the Platinum Group Metals, December 1968.

Rosenblum, Myron, Chemistry of the Iron Group Metallocenes: Ferrocene, Ruthenocene, and Osmocene, Part 1, John Wiley and Sons (New York) (1965).

Rylander, Paul N., Catalytic Hydrogenation Over Platinum Metals, Academic Press (New York) (1967).

Rylander, Paul N., and Duane R. Steele, "Osmium Catalyzed Hydrogenations," Engelhard Ind. Tech. Bulletin, **10** (1), 17 (1969).

Rylander, Paul N., "Osmium Tetroxide as an Oxidation Catalyst in Organic Synthesis," Englehard Ind. Tech. Bulletin, **9** (3), 90 (1968).

Ryosuke, Goto, "Industrial Purification Process of Osmium," Electro-Chemical Soc. of Japan, **22**, 5 (1954).

Samsahl, K., and D. Brune, "Simultaneous Determination of 30 Trace Elements in Cancerous and Noncancerous Human Tissue Samples by Neutron Activation Analysis," Intern. J. Appl. Radiation Isotopes, **16**, 273 (1965).

Sax, Irving N., Dangerous Properties of Industrial Materials, 2nd Edition, Reinhold Publishing Company,(New York) (1963).

Schiechl, Von H., "Der Chemismus der OsO_4-Fixierung and Einfluss auf die Zellstrucktur," Acta Histochemica, **10-11**, Supplement, 165-171 (1971).

Schlatter, C. H., and I. Schlatter-Lanze, "A Simple Method for the Regeneration of Used Osmium Tetroxide Fixative Solutions," J. Microscopy, **94** (1), 85-87 (1971).

Shelef, M., and H. S. Gandhi, "Ammonia Formation in Catalytic Reduction of Nitric Oxide by Molecular Hydrogen," Ind. Eng. Chem. Prod. Res. Develop., 11 (4), 393 (1972).

Somers, E., "Plant Pathology: Fungitoxicity of Metal Ions," Nature, 184 (4684), 475 (1959).

Stokinger, H. E., "Chapter XXVII - The Metals (Excluding Lead)" in Industrial Hygiene and Toxicology, Vol. II, 2nd Revised Edition, Frank A. Patty, Ed., Interscience Publishers (New York) (1958).

Strangeways, T. S. P., and R. C. Canti, Quart. J. Micr. Sci., 71, 1 (1927).

Swaine, D. J., The Trace-Element Content of Fertilizers, Commonwealth Agri. Bureau, Farmham Royal (Bucks, England) (1962).

Taylor, H., and F. E. Beamish, "Isolation of Osmium and Ruthenium by Ion-Exchange Paper and Subsequent Determination by X-Ray Fluorescence," Talanta, 15, 497 (1968).

Taylor, S. R., "Abundance of Chemical Elements in the Continental Crust: A New Table," Geochim. Cosmochim. Acta, 28, 1273 (1964).

Tennant, S., Phil Trans., 94, 411 (1804).

Tipton, Isabel H., "Trace Elements in Human Tissue," Am. J. Clinical Nutrition, 23 (2), 123-125 (1970).

Tugwell, Gilbert L., "Industrial Applications for the Noble Metals," Metal Progress, 73-78, July 1965.

U.S. Department of the Interior, Bureau of Mines, "Spark-Source Mass Spectrometer Investigation of Coal Particles and Coal Ash," U.S. Bureau of Mines Progress Report No. 42 (1971).

Van Loon, J. C., and F. E. Beamish, "A Fire Assay for Osmium in Sulfide Concentrates," Analytical Chemistry, 36 (4), 372 (1964).

Vernadsky, V. I., Supplement to Handbuch der bestimmenden Mineralogie (original publication Vienna, 1845), p. 752 (1914).

Vines, Raymond F., "Precious Metal Electrodeposits for Electrical Contact Service," Plating, 59, 923-925 (1967).

von Reis, G., and A. Swensson, Acta Med. Scand. Suppl., 259, 27 (1951).

Ware, Glen C., "Platinum-Group Metals" in Mineral Facts and Problems, Bureau of Mines Bulletin 630, U.S. Department of the Interior, U.S. Government Printing Office (Washington,D.C.), pp. 711-719 (1965).

Westerfield, W. W., "Effect of Metal-Binding Agents on Metalloproteins," Federation Proceedings, Supplement 10, 20 (3), 158-176 (1961).

Wigglesworth, V. G., Proc. Roy. Soc., B147, 185 (1967).

Wright, T. L., and Michael Fleischer, "Geochemistry of the Platinum Metals," Geo. Survey Bulletin, 1214-A (1965).

SUBJECT INDEX

airborne palladium 93
Alaska
  osmium metals in 147
  platinum metals in 7, 8, 43
alloys
  of osmium 143, 147, 163
  of palladium 7, 27, 28, 33, 51, 62, 78, 79, 112, 113
analytical methods
  for palladium 115-118

Canada
  osmium metals in 147, 150
  platinum metals in 3, 7, 8, 10, 11, 45
carcinogenicity
  of palladium 5, 53, 57
catalysts
  osmium in 143, 156, 158, 159, 166
  palladium in 25, 26, 32, 34-42, 49, 50, 85-91
chemical industry, palladium catalysts in 34, 35, 47-49
  osmium uses 143, 158, 162
chemistry
  of osmium 169-174
  of palladium 105-114
cladding
  palladium 31
coatings
  palladium 29-31
copper ores and scrap
  osmium in 144, 147, 151, 165, 166
  palladium in 3, 7, 8, 12, 13, 20, 44
corrosion
  of palladium 106-109
crustal abundance
  of palladium 6

dental uses of palladium 33, 51, 80, 81

electronics
  electronic contacts and switches 33, 50, 74, 75, 143, 158, 163
  fuel cells 77
  scrap 9, 24, 25
  telephone equipment 33
  thermocouples 76, 77
  vacuum tubes 163
  windings and resistors 76
environmental losses
  of osmium 143, 144, 157
  of palladium 4, 5, 43-51, 57
exports
  of palladium 10
  of platinum group metals 69

fish
  toxicity of palladium to 54

general services administration
  purchases of osmium 159
  stocks of palladium 67
gold ore
  osmium in 147
  palladium in 3, 8, 12, 43
growth stimulation
  by palladium 55

human health hazards
  of osmium 144, 145, 163, 167, 168
  of palladium 5, 56, 57
humans
  effects of osmium on 144, 145, 163, 167, 168
  effects of palladium on 52, 53

humans (concluded)
  effects of sulfuric acid on 37
  palladium toxicity 5, 52, 56, 57
hydrogenation catalyst
  palladium as 34-36

imports
  osmium 155
  palladium 9, 64, 65
  platinum-group metals 9

lead ore
  palladium in 3, 8, 18, 19

medical industry
  osmium uses 143, 161, 162
metal scrap 3, 8, 9
mining
  osmium 150-152
  palladium 12, 13, 46
mufflers
  environmental loss of palladium from 49, 50
  use of palladium catalyst in 4, 5, 33, 36-42, 51, 92

nickel ore
  osmium in 147, 151
  palladium in 7, 13-18, 44
nuclear waste
  palladium in 27

occurrence of osmium
  in catalysts 156
  in fixative solutions 156
  in natural sources 146, 149, 164
  in the environment 146, 149, 164-167
occurrence of palladium
  in catalysts 25, 26, 34-42, 85-97
  in metal scrap 3, 8, 20-27
  in natural compounds 63
  in natural sources 7, 8, 63
  in nuclear waste 27
  in the environment 3, 5, 6, 43-51
  in various sources 61
osmium
  uses 143
oxidation
  of osmium 143
  of palladium 46, 109, 110

palladium
  analysis 115-118
  chemistry 105-114
  physical properties 94-104
  uses 4, 32-43, 74-84
petroleum refining, palladium catalysts in 36, 47-49
pharmaceutical industry, palladium catalysts in 35
physical properties
  of osmium and its compounds 175-178
  of palladium and its compounds 94-104
plants
  effects of palladium on 55
platinum-group metals 3, 7, 9, 146, 147, 150, 157
processing
  palladium ore and scrap 12-31
production
  of osmium in U.S. 152, 154
  of palladium 3, 9, 12-31, 66
production facilities
  palladium 72, 73

refining
  osmium 150-155
  palladium 13-19

sales
  of osmium 143, 144, 159
  of palladium 4, 10, 70
secondary recovery
  of osmium 156, 157
  of palladium 3, 20-27
smelting
  environmental losses of osmium from 165
  environmental losses of palladium from 44, 46, 47
sources of osmium
  (see occurrences of osmium)
sources of palladium
  (see occurrence of palladium)
South Africa
  osmium metals in 147, 150
  platinum metals in 3, 7, 8, 10, 11
stocks
  of palladium 9, 67, 68
sulfuric acid
  from catalytic mufflers 37
supply and demand
  osmium 144, 146
  palladium 71

therapeutic effects and uses
  osmium compounds 162
  palladium compounds 52, 53, 57, 81, 82
tissue staining 143, 159, 161, 162, 166
toxicity
  of osmium 145, 167
  of palladium 5, 52-55, 56, 57

United States
  osmium metals in 147, 150
  platinum metals in 7, 8, 12, 43
uses
  of osmium 143, 158-164
  of palladium 4, 10, 25, 32-42, 74-92
USSR
  osmium metals in 147
  platinum metals in 3, 7, 8, 10, 11

volatilization
  of osmium 144
  of palladium 46, 47, 109, 110